T0344465

A Robotic Framework for the Mobile Manipulator

By proposing and forming a mobile manipulator for modern multifloor buildings, *A Robotic Framework for the Mobile Manipulator: Theory and Application* helps readers visualize an end-to-end workflow for making a robot system work in a targeted environment. From a product-oriented viewpoint, this book is considered as a bridge from theories to real products, in which robotic software modules and the robotic system integration are mainly concerned. In the end, readers will have an overview of how to build and integrate various single robotic modules to execute a list of designed tasks in the real world, as well as how to make a robot system work independently, without human interventions. With references and execution guidelines provided at the end of each chapter, this book will be a useful tool for developers and researchers looking to expand their knowledge about robotics and robotic software.

Chapman & Hall/CRC Artificial Intelligence and Robotics Series

Series Editor: Roman Yampolskiy

Cunning Machines
Your Pocket Guide to the World of Artificial Intelligence
Jędrzej Osiński

Autonomous Driving and Advanced Driver-Assistance Systems
Applications, Development, Legal Issues, and Testing
Edited by Lentin Joseph, Amit Kumar Mondal

Digital Afterlife and the Spiritual Realm
Maggi Savin-Baden

A First Course in Aerial Robots and Drones
Yasmina Bestaoui Sebbane

AI by Design
A Plan for Living with Artificial Intelligence
Catriona Campbell

The Global Politics of Artificial Intelligence
Edited by Maurizio Tinnirello

Unity in Embedded System Design and Robotics
A Step-by-Step Guide
Ata Jahangir Moshayedi, Amin Kolahdooz, Liao Liefa

Meaningful Futures with Robots
Designing a New Coexistence
Edited by Judith Dörrenbächer, Marc Hassenzahl, Robin Neuhaus, Ronda
Ringfort-Felner

Topological Dynamics in Metamodel Discovery with Artificial Intelligence
From Biomedical to Cosmological Technologies
Ariel Fernández

A Robotic Framework for the Mobile Manipulator
Theory and Application
Nguyen Van Toan and Phan Bui Khoi

AI in and for Africa
A Humanist Perspective
Susan Brokensha, Eduan Kotzé, Burgert A. Senekal

For more information about this series please visit:
https://www.routledge.com/Chapman–HallCRC-Artificial-Intelligence-
and-Robotics-Series/book-series/ARTILRO

A Robotic Framework for the Mobile Manipulator
Theory and Application

Nguyen Van Toan
Phan Bui Khoi

CRC Press
Taylor & Francis Group
Boca Raton London New York

CRC Press is an imprint of the
Taylor & Francis Group, an **informa** business

First edition published 2023
by CRC Press
6000 Broken Sound Parkway NW, Suite 300, Boca Raton, FL 33487-2742

and by CRC Press
4 Park Square, Milton Park, Abingdon, Oxon, OX14 4RN

CRC Press is an imprint of Taylor & Francis Group, LLC

© 2023 Nguyen Van Toan and Phan Bui Khoi

Library of Congress Cataloging-in-Publication Data

Names: Nguyen, Van Toan (Robotics engineer), editor. | Phan, Bui Khoi, editor.
Title: A robotic framework for the mobile manipulator : theory and
   application / Nguyen Van Toan, Phan Bui Khoi.
Description: First edition. | Boca Raton, FL : CRC Press, 2023. | Series:
   Chapman & Hall/CRC artificial intelligence and robotics series |
   Includes bibliographical references and index.
Identifiers: LCCN 2022052385 (print) | LCCN 2022052386 (ebook) | ISBN
   9781032403083 (pbk) | ISBN 9781032392608 (hbk) | ISBN 9781003352426 (ebk)
Subjects: LCSH: Autonomous robots. | Mobile robots. | Manipulators (Mechanism)
Classification: LCC TJ211.415 .R59 2023 (print) | LCC TJ211.415 (ebook) |
   DDC 629.8/932--dc23/eng/20221230
LC record available at https://lccn.loc.gov/2022052385
LC ebook record available at https://lccn.loc.gov/2022052386

ISBN: 978-1-032-39260-8 (hbk)
ISBN: 978-1-032-40308-3 (pbk)
ISBN: 978-1-003-35242-6 (ebk)

DOI: 10.1201/9781003352426

Typeset in Times
by KnowledgeWorks Global Ltd.

Contents

Preface

By proposing and forming a mobile manipulator for modern multi-floor buildings, this book helps readers visualize an end-to-end workflow for making a robot system work in a targeted environment. From a product-oriented viewpoint, this book is considered as a bridge from theories to real products, in which robotic software modules and robotic system integration are mainly concerned. In the end, readers will have an overview of how to build and integrate various single robotic modules to serve a list of designed tasks in the real world. One thing that needs to be emphasized in this book is how to make a robot system work independently, without human interventions. It is known that a robot system is constructed by various types of single functions. To make a robotic product work independently, the robot system must have the capability to switch its missions automatically. However, almost all studies have just focused on single robotic software modules, although the robotic system integration is crucial. For an overview on this issue, this book presents the following contents: in Chapter 1, the purpose of this book is clarified by presenting the importance and current issues in the robot system integration, as well as works to be done for the stated goal. In Chapter 2, a task-oriented approach is used to propose hardware and software modules for a robot system. For illustration, a robot system is proposed to work in modern multi-floor buildings. After that, a robotic framework is constructed to help the proposed robot system to change its working floors. Next, the kinematics and dynamics of the proposed robot system are analyzed in Chapter 3. In Chapter 4, core features of mobile robot navigation are conducted, including localization and mapping, global positioning, environment modeling, path generation, and following with collision avoidance. In Chapter 5, the manipulator manipulation work of the proposed robot system is briefly presented to adapt to software system, regarding the formulation of manipulator modeling, trajectory planning, and control. In Chapter 6, some perception works are presented to help the proposed mobile manipulator to change its working floor successfully, including elevator button detection, elevator button light-up status checking, elevator door status checking, human detection, and floor number recognition. Finally, in Chapter 7, a robotic decision-making system is presented to integrate single robotic modules, whose purpose is to help the proposed robot system switch its missions automatically and work without human interventions.

It is noted that the proposed framework is started right after the robot is requested to change its work-floor and then finished after it entered the targeted floor successfully. This proposed framework is used as a specific example to help readers imagine about real-life applications of the robot system more easily. Although changing floors is the only part of the work list of the proposed robot system in modern multi-floor buildings, it brings an overview of how to make a robot system serve a specific task in the real world without loss of generality. Of course, in reality, the proposed robot system must conduct many other desired tasks in modern multi-floor buildings, such as the map switching to help the robot system adapt new working floors, auto-docking, charge dock (battery auto charging), and safety operations. These mentioned tasks can be straightforwardly added to the expandable framework in this book.

By virtue of its flexibility, the mobile manipulator is an interesting type of robotics, which offers many potential applications to reduce human labor. This is one of the reasons why the mobile manipulator is chosen to become the representative subject of this book. Nevertheless, authors believe that this book is introduced to become the reference for robotic students, developers, and researchers not only to the mobile manipulator but also other kinds of robots.

Nguyen Van Toan

About the Authors

Nguyen Van Toan is the Principle Researcher at Robotics R&D Center, Syscon, Robot Land, South Korea. He is currently completing his Doctor of Philosophy at the Department of Electrical and Information Engineering, Seoul National University of Science and Technology, Seoul, South Korea.

Phan Bui Khoi received his Doctorate degree in Robotics from the Mechanical Engineering Research Institute of the Russian Academy of Sciences in 1997. He is an Associate Professor of dynamics and control of robot and mechatronic system at Hanoi University of Science and Technology.

Introduction

1

Nguyen Van Toan and Phan Bui Khoi

Robotic applications have been increasingly popular in human activities. It is clear that the development of robotic technology will be accelerated more strongly if its research approaches are simpler and easier to apply in practice. From a product-oriented viewpoint, operating a robot with a low level of automation is really challenging for users, especially users without deep knowledge of robotics and systems. Therefore, making a flexible robot system that can easily respond to changes in its targets and working environments is a crucial issue. If the robot can complete its missions at a high or full level of automation, human labor and human errors will be dramatically reduced. To do this, a robotic decision-making system must be conducted so that the robot can work without deep interventions of operators [1–4]. Such robot system integration brings significant benefits: (1A1) speed and consistency of executions of tasks are increased, (1A2) accuracy and precision of robot manipulations are improved, (1A3) human labor and costs are reduced, and (1A4) a safe working environment is created since human errors are reduced with simple or no required operations. However, the system integration task is usually underestimated in traditional designs. Almost all research works have just focused on single robotic modules while unexpected and unforeseen issues have usually been raised during the integration process, regarding interorganizations and inter-subsystems. As presented in Ref. [4], a lack of interoperability and compatibility causes inaccurate executions of robotic products. For example, subsystems do not work together, the integration environment is inadequate, and the robotic product is delivered without all of its components been tested.

To construct a robotic product and integrate its all single modules concisely and efficiently, the prerequisite work is to discover its desired applications and working environments since a robot system can only serve a set of

DOI: 10.1201/9781003352426-1

specific tasks in specific environments. It is clear that benefits are only maximized by allocating resources appropriately, in the right places for the right tasks regardless of human or robot. For illustration, a person who can only communicate in Vietnamese and Korean is assigned to interpret Vietnamese-English, but not Vietnamese-Korean. It must have been a terrible assignment from the manager. Similarly, the task of grasp-and-place is impossible to complete if it is assigned to a mobile robot, but not a manipulator or a mobile manipulator. On the other hand, duplicate modules created for similar robotic tasks should be redundant and resource-wasting. Thus, the first important step is to identify the problems to be solved and the working environment of the targeted robot system. Based on that information, robotic modules and integration methods can be decided properly.

Currently, the mankind has had a huge treasure of knowledge about science and technology in general, as well as about robotics in particular. First, to develop a product in any field, people must delve deeply into related scientific and technological issues. On that basis, methodologies and implementation approaches will be proposed to match their goals. In robotics, with a great amount of available publications (research and technical papers, and books), robotists can learn, exploit, and even create their own robots. For example, Refs. [5–18] present most of the areas related to robotics, while Refs. [19–27] present robot-designing problems. Those publications delved into individual topics about robotics. However, it should be very challenging for beginners or even experienced developers since the study of various documents with thousands of pages consumes a considerable amount of their time and energy. By the time, publications are constantly being updated while a lot of knowledge is not really necessary for their goals and their specific products. In fact, very few concise documents present fully and thoroughly an end-to-end process for making a robotic product in targeted environments. To provide such kind of reference, this book presents an end-to-end workflow for making a robot system to work independently in a targeted environment, without human interventions. In the beginning, a robot system is assumed to have been ordered by customers to serve some applications in modern multi-floor buildings, which consists of an autonomous mobile robot (the differential drive model) and an eye-on-hand manipulator (6 degrees of freedom [6-DOF]). Then, a robotic framework for the proposed mobile manipulator to change its work floors is used as a specific example to demonstrate real-life applications of a robot system. This book presents only knowledge of the core features of the targeted product and encourages readers learn more details in given references and resources. Thus, with a not-too-large content focusing on a typical robot system, this book gives readers an overview and factual observations.

Service robots have become more and more common to provide assistances to persons in both daily life and manufacturing. In recent decades, the use of such kinds of robots in modern multi-floor buildings has received a lot of attention, in which the elevator operation is one of the tough challenges for the robot system because it has to serve different tasks in different floors. To operate the elevator, two methods can be considered: (1B1) enable the communication between the robot system and the elevator system via the wireless connection, or (1B2) equip the robot system with the capacity to detect and manipulate elevator components such as elevator buttons, the elevator door, or the floor number display. By using (1B1), the robot system controls the elevator system by sending commands via the wireless connection when the robot is requested to move upward or downward to another floor. However, this method is not setting-independent, normally time-consuming and complicated. By contrast, (1B2) is known as a human-like method and independent on the elevator system. The method (1B2) is therefore more interesting and potential than the method (1B1). With its self-driving and manipulation capabilities, the mobile manipulator is currently judged to be the most compatible with the method (1B2) to work in such working environment. Besides, the mobile manipulator also provides many other applications, such as handling-pick and place, loading and positioning, machinery machining and assembly, inspection, and testing. Some of the advantages of mobile manipulators are as follows [28]: (1C1) high versability due to integrating the manipulator and the mobile base with self-driving navigation ability (free navigation), (1C2) high efficiency and flexibility in tasks, and (1C3) collaborative workspace, because the mobile manipulator is completely autonomous. It can work in different locations and share workspaces and tasks with humans and other robots, (1C4) perform tasks with human-like capacities in a variety of environments, and (1C5) the gap between teleoperation and full autonomy is bridged. These advantages help to improve the production process and increase the profitability. These are the reasons why a mobile manipulator in modern multi-floor buildings is chosen to become the representative subject of this book. Along with mentioned benefits, some challenges also arose [29]: (1D1) kinematics and dynamics are more difficult to analyze and synchronize, (1D2) control and system integration are more complex, and (1D3) more requirements in designs, such as the mobile manipulator must feature some attributes: lightweight, compact, power-efficient, power-dense, portable, and be rugged to withstand weather, shock, and vibration. These challenges will open up many interesting research topics in the future. Regarding the proposed robot system in modern multi-floor buildings, the structure of the robot arm depends on technical requirements of manipulations. To go up and down other floors, the elevator operation is conducted by detecting and manipulating elevator components, such as elevator buttons, the elevator door, or the floor number display. With a suitable pose of the mobile base, mentioned detection and manipulation works can be completed by using 3-DOF manipulator. However, the poses of the mobile robot are not always accurate. Besides, the manipulator is also used for other tasks. Therefore, it is proposed with 6-DOF to fully guarantee the poses of the end effector within the working space of the robot arm. Although the form of the manipulator can be open chain, closed chain, or parallel, it is formed as an open chain in order to make the presentation of this book as well as to help the readers absorb most easily. On that basis, readers can continue to develop other structures of manipulators. For the same reason, the differential drive model is chosen for the autonomous mobile robot, but not omni-wheels.

Back to the modern multi-floor buildings, a clear reality is that a mobile manipulator usually serves many different tasks in such environment. It is not possible here to present all of them in detail. Without loss of generality, to elucidate the final goal of this book (the robot system integration), the scheme for changing work floors of the mobile manipulator is executed since it covers almost all modules of the proposed robot system such as the robotic navigation, manipulator manipulation, robot perception, decision-making, and integration. Besides, the elevator operation is also an interesting topic and is a prerequisite ability to enable mobile manipulators to work in modern multi-floor buildings since they have no capacity to climb the stairs. Based on our observations, to help the robot system change its working floors successfully by using method (1B2), some sets of robotic modules should be considered [30–33]: (1E1) the mobile robot navigation (self-driving task) to help the robot system travel in its workspace; (1E2) perception works, including elevator button detection, button light-up status recognition, elevator door status checking, human detection, and floor number recognition; (1E3) the manipulator manipulation, such as pressing elevator buttons, or changing the camera view; and (1E4) the integration of (1E1), (1E2), and (1E3) should be conducted to make the robot system work at the high level of the automation. Single robotic modules and the system integration are then conducted on the robot operating system 2 (ROS2) and presented in remaining chapters as follows: in Chapter 2, a task-oriented approach is used to propose hardware and software modules for a robot system. For illustration, a robot system is proposed to work in modern multi-floor buildings. After that, a robotic framework is constructed to help the proposed robot system to change its working floors. Next, the kinematics and dynamics of the proposed robot system is analyzed in Chapter 3. In Chapter 4, core features of mobile robot navigation are conducted, including localization and mapping, global positioning, environment modeling, path generation, and following with collision avoidance. In Chapter 5, the manipulator manipulation work of the proposed robot system is briefly presented to adapt software system, regarding the formulation of the manipulator modeling, the trajectory planning, and control. In Chapter 6, some perception works are presented to help the proposed mobile manipulator to change its working floors successfully, including elevator button detection, elevator button light-up status checking, elevator door status checking, human detection, and floor number recognition. Finally, in Chapter 7, a robotic decision-making system is presented to integrate single robotic modules, whose purpose is to help the proposed robot system switch its missions automatically and work without human interventions. In inheritance, other tasks can be straightforwardly added to the presented framework.

ABBREVIATIONS

DOF: Degrees of Freedom

REFERENCES

[1] Ryo Hanai, Kensuke Harada, Isao Hara, Noriaki Ando. *Design of robot programming software for the systematic reuse of teaching data including environment model.* ROBOMECH Journal, vol. 5, no. 21, pp. 1–16, 2018. https://doi.org/10.1186/s40648-018-0120-z

[2] Haiyan Wu, Rikke Bateman, Xinxin Zhang, Morten Lind. *Functional modeling for monitoring of robotic system.* Applied Artificial Intelligence, vol. 32, no. 3, pp. 229–252, 2018.

[3] Stefan Praschl, Michael Fung. *Mobile robotics: manufacturing and engineering technology.* Technical Description, WSC2021_TD23_EN, WorldSkills, 22 September 2020.

[4] Mary Beth Chrissis, Mike Konrad, Sandy Shrum. *CMMI: guidelines for process integration and product improvement.* 2nd edn., Addison-Wesley, Boston, 2007.

[5] John J. Craig. *Introduction to robotics: mechanics and control.* 3rd edn., Pearson, London, 2004.

[6] Lung-Wen Tsai. *Robot analysis: the mechanics of serial and parallel manipulators.* John Willey & Sons, New York, 1999.

[7] Bruno Siciliano, Lorenzo Sciavicco, Luigi Villani, Giuseppe Oriolo. *Robotics: modelling, planning and control.* Springer-Verlag, London, 2009.

[8] Bruno Sicilianno, Oussama Khatib. *Springer handbook of robotics.* Springer-Verlag, Berlin-Heidelberg, 2016.

[9] Mark W. Spong, Seth Hutchinson, Mathukumalli Vidyasagar. *Robot modeling and control. 1st edn.* Wiley, Hoboken, 2005.

[10] Christoph Bartneck, Tony Belpaeme, Friederike Eyssel, Takayuki Kanda, Merel Keijsers, Selma Sabanovic. *Human–robot interaction: an introduction.* 1st edn., Cambridge University Press, Cambridge, 2020.

[11] Lentin Joseph. *Mastering ROS for robotics programming: design, build, and simulate complex robots using the robot operating system.* 3rd edn., Packt Publishing Ltd., Birmingham, 2021.

[12] Luc Jaulin. *Mobile robotics.* ISTE Press—Elsevier, London, 2016.

[13] Antoni Grau, Yannick Morel, Ana Puig-Pey, Francesca Cecchi. *Advances in robotics research: from lab to market. Subtitle: ECHORD++: robotic science supporting innovation.* Springer, Cham, 2020.

[14] Peter Corke. *Robotics and control. Subtitle: Fundamental algorithms in MATLAB.* Springer, Cham, 2022.

[15] Jose M. Pardos-Gotor. *Screw theory in robotics: an illustrated and practicable introduction to modern mechanics.* 1st edn., CRC Press, Boca Raton, 2022.

[16] Jaime Gallardo-Alvarado, José Gallardo-Razo. *Mechanisms: kinematic analysis and applications in robotics.* 1st edn., Academic Press: An imprint of Elsevier, Cambridge, 2022.

[17] Andrii Kudriashov, Tomasz Buratowski, Mariusz Giergiel, Piotr Małka. *SLAM techniques application for mobile robot in rough terrain.* Springer, Cham, 2020.

[18] Saïd Zeghloul, Med Amine Laribi, Marc Arsicault. *Mechanism design for robotics: MEDER 2021.* Springer, Cham, 2021.

[19] David R. Shircliff. *Build a remote controlled robot.* 1st edn., McGraw-Hill/ TAB Electronics, New York, 2002.

[20] Gordon McComb. *Robot builder's sourcebook: over 2500 sources for robot parts.* 1st edn., McGraw Hill/TAB Electronics, New York, 2002.

[21] Gordon McComb. *Robot builder's bonanza.* 4th edn., McGraw-Hill/TAB Electronics, New York, 2011.

[22] Karl Williams. *Amphibionics: build your own biologically inspired reptilian robot.* 1st edn., McGraw-Hill/TAB Electronics, New York, 2003.

[23] Owen Bishop. *Robot builder's cookbook: build and design your own robots.* 1st edn., Newnes, London, 2007.

[24] Pete Miles, Tom Carroll. *Build your own combat robot.* McGraw-Hill/Osborne Media, New York, 2002.

[25] Giacomo Marani, Junku Yuh. *Introduction to autonomous manipulation: case study with an underwater robot, SAUVIM.* Springer, Berlin-Heidelberg, 2014.

[26] Cameron Hughes, Tracey Hughes. *Robot programming: a guide to controlling autonomous robots.* 1st edn., Que Publishing, New York, 2016.

[27] Eugene Kagan, Nir Shvalb, Irad Ben-Gal. *Autonomous mobile robots and multi-robot systems: motion-planning, communication, and swarming.* Wiley, Hoboken, 2020.

[28] Mads Hvilshoj, Simon Bogh. *"Little Helper"—An autonomous industrial mobile manipulator concept.* International Journal of Advanced Robotic Systems, vol. 8, no. 2, 2011. https://doi.org/10.5772/10579

[29] Martin Sereinig, Wolfgang Werth, Lisa-Marie Faller. *A review of the challenges in mobile manipulation: systems design and RoboCup challenges.* Elektrotechnik und Informationstechnik, vol. 137, pp. 297–308, 2020.

[30] Nguyen Van Toan, Jeong Jin-Hyeon, Jaewon Jo. *An efficient approach for the elevator button manipulation using the visual-based self-driving mobile manipulator.* Industrial Robot: The International Journal of Robotics Research and Application, 2022. https://doi.org/10.1108/IR-03-2022-0063

[31] Delong Zhu, Zhe Min, Tong Zhou, Tingguang Li, Max Q.-H. Meng. *An Autonomous eye-in-hand robotic system for elevator button operation based on deep recognition network.* IEEE Transactions on Instrumentation and Measurement, *vol.* 70, pp. 1–13, 2021, Art no. 2504113.

[32] Kerstin Thurow, Lei Zhang, Hui Liu, Steffen Junginger, Norbert Stoll, Jiahao Huang. *Multi-floor laboratory transportation technologies based on intelligent mobile robots.* Transportation Safety and Environment, vol. 1, no. 1, pp. 37–53, 2019.

[33] Ali A. Abdulla, Mohammed M. Ali, Norbert Stoll, Kerstin Thurow. *Integration of navigation, vision, and arm manipulation towards elevator operation for laboratory transportation system using mobile robots.* Journal of Automation, Mobile Robotics & Intelligent Systems, vol. 11, no. 4, pp. 34–50, 2017.

Task-Oriented Robot System Proposal

2

Nguyen Van Toan

In this chapter, a task-oriented approach is used to propose hardware and software modules for a robot system. For illustration, a robot system is proposed to work in modern multi-floor buildings. After that, a robotic framework is constructed to help the proposed robot system to change its working floors. The proposed robot system, the constructed framework for changing the robot's working floors, and its required hardware and software modules are then used throughout other chapters in this book.

TASK-ORIENTED ANALYSIS

To propose hardware and software modules for a robot system, the prerequisite work is to discover its desired applications and working environments since a robot system can only serve a set of specific tasks in specific environments. In other words, the task list should be the root of a robot system, which is normally constructed as the form of a tree. Basically, this tree consists of various robotic tasks; each task consists of various missions, and each mission consists of various actions, as presented in Figure 2.1.

It is noted that the number of missions in a specific task (as well as the number of actions in a specific mission) is variable, depending on how many jobs must be conducted to complete the task (or mission) and how big a mission

DOI: 10.1201/9781003352426-2

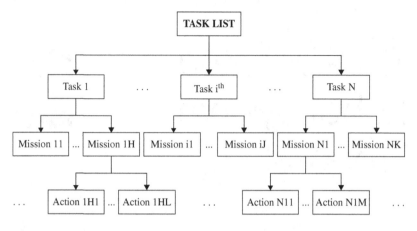

FIGURE 2.1   The tree of the robotic task list.

(Photograph by Nguyen Van Toan.)

(or an action) is defined. For illustration (called Example 2.1), the pallet delivery is considered as a task of an autonomous mobile robot (AMR) in the factory, in which some missions can be defined as (2A1) the pallet pick-up, (2A2) traveling to the targeted point (in front of a specific working station), and (2A3) the pallet place-off. Besides, each mission consists of some actions. For example, the mission (2A1) is completed by actions: dock detection, obstacle detection, docking-in, pallet lift-up, docking-out, and pallet lift-down. Another illustration (called Example 2.2) is the auto-charging task of the AMR in the factory. This task is defined by some missions as (2B1) checking the current battery status, (2B2) traveling to the targeted point (in front of a specific charge-dock), and (2B3) the auto charging. Similarly, each mission includes a number of actions. For example, some actions can be defined in mission (2B3) as charge dock detection, obstacle detection, docking-in, and docking-out. Here, the mission (2B2) is conducted if the current battery is less than 30 percent. The targeted point in this mission (2B2) is selected from among idle chargers at the present time. Besides, the docking-out action in mission (2B3) is implemented if the current battery is bigger than 95 percent. As presented in Examples 2.1 and 2.2, an action is normally defined as a software module, a mission consists of at least one action, and a task consists of at least one mission. Moreover, missions in different tasks (or actions in different missions) can be duplicated.

After the task list of a robot system is revealed, its hardware and software modules can be then envisioned. As in Examples 2.1 and 2.2, the AMR must be equipped with some software modules such as mapping, localization, navigation, obstacle avoidance (both 2D and 3D obstacles), dock detection (L shape,

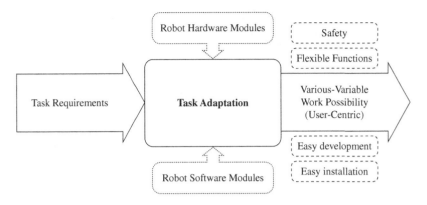

FIGURE 2.2   The task flow of a robot system.
(Photograph by Nguyen Van Toan.)

V shape, or some other kinds of docks), lifting functions, the emergency safety, and so on. For these software modules, the robot system must be constructed by compatible hardware modules such as actuated wheels, castors, LiDAR sensors, 3D cameras, lifting mechanisms, and so on. It is reminded that those proposed robot hardware and software modules are not limited to desired tasks in Examples 2.1 and 2.2. They can be used for other tasks or even being possibly extended for potential tasks in the future. In general, the choice of software modules, hardware modules, and the robot system integration should follow some standards, such as safety, function flexibility, easy development, and easy installation. By satisfying these requirements, a proposed robot hardware module (or a proposed robot software module) can be simultaneously used for various purposes (tasks) and also has the ability to adapt future applications. Of course, concerns about standards can be changed over time. In this book, the software integration is mainly emphasized. To reduce the cost of the robot system, software modules should be developed so that the number of used hardware modules is the least. The task flow of a robot system is presented in Figure 2.2.

A ROBOT SYSTEM IN MODERN MULTI-FLOOR BUILDINGS

As presented, robotic hardware and software modules must be decided through the desired applications of the targeted robot system and its working environment. In this section, a robot system is assumed to have been ordered

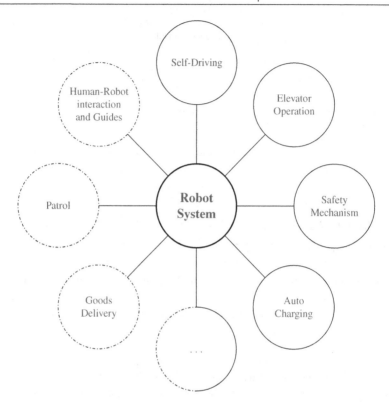

FIGURE 2.3   Basic modules of a robot system in modern multi-floor buildings. (Photograph by Nguyen Van Toan.)

by customers to serve some applications in modern multi-floor buildings, such as goods delivery and patrol. Such an assumption is considered since service robots have become more and more common to provide assistances to persons in both daily life and manufacturing. Moreover, the use of service robots in modern multi-floor buildings has received great attention in the recent decade. Therefore, a robot system in such environment is used as the representative subject of this book. Some basic modules of the targeted robot system are mentioned in Figure 2.3. Here, solid circles represent core capabilities of the robot system. Dashed circles represent desired applications, depending on customers' requests. Each dashed circle is constructed by a set of solid circles. A solid or dashed circle can be defined as an individual task of the robot system. Their missions and actions can be duplicated as presented in the previous section, or even a task can be a child of another task (parent task). For example, *"Elevator Operation," "Self-Driving,"* and *"Safety*

Mechanism" must be children of *"Goods Delivery."* It is reminded that future tasks can be adaptively added to the robot system, as represented by the circle (a half solid and a half dashed) in Figure 2.3.

The targeted robot system as shown in Figure 2.3 covers various areas in robotics such as robotic navigation, manipulator manipulation, robot perception, decision-making, and system integration. For illustration, to ensure self-driving ability, the robot system must be designed as an AMR, which includes some basic functions such as mapping, localization, and collision-free navigation. To operate the elevator, two methods can be considered: (2C1) enable the communication between the robot system and the elevator system via the wireless connection, or (2C2) equip the robot system with the capacity to detect and manipulate elevator components such as elevator buttons, the elevator door, or the floor number display. By using (2C1), the robot system controls the elevator system by sending commands via the wireless connection when the robot is requested to move upward or downward to another floor. However, this method is not setting-independent, normally time-consuming, and complicated. By contrast, (2C2) is known as a human-like method, and independent on the elevator system. The method (2C2) is therefore considered for the robot system in this book. To conduct the method (2C2), the robot system must be able to detect and manipulate elevator buttons, check the button light status, check the elevator door status, recognize the appearance of persons, and recognize the targeted floor number. These abilities can be achieved by using a manipulator with a camera mounted on its end effector. This eye-on-hand method is also used for other tasks such as goods delivery, patrol, and human-robot interaction as mentioned in Figure 2.3. Besides, the auto-charging task is completed by conducting three modules: the battery status checking, the dock detection, and the docking planning. Here, the AR marker-based methods (using the RGB-D camera) and V-maker/L-marker-based methods (using the LiDAR sensor) are potential choices for the dock detection. Finally, the safety mechanism should be qualified by the 2D/3D obstacle avoidance and the emergency brake. Now, the proposed robot system can be imagined as a mobile manipulator, including an AMR and an eye-on-hand manipulator (placed on the AMR). Basic modules of the proposed robot system and their roles are described in Figure 2.4.

In Figure 2.4, the AMR is driven by two actuated wheels and supported by four casters. Two RGB-D cameras are installed in the front of the AMR. Their merged field of view (FOV) is around 180 horizontal degrees and 58 vertical degrees. In addition, two LiDAR sensors (one is installed in the left corner of the rear side and another is installed in the right corner of the front side) are used. Their merge scan view is 360 horizontal degrees, in which the AMR body is eliminated. This AMR takes place in the navigation work. Besides, a manipulator is located on the mobile robot. One RGB-D camera

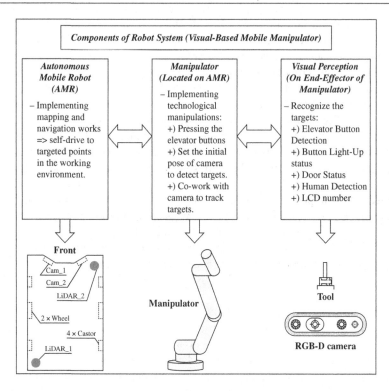

FIGURE 2.4   A description of the proposed robot system.

(Photograph by Nguyen Van Toan.)

and one tool are mounted on the manipulator's end effector, whose purpose is to detect and manipulate designed targets. Here, the manipulator Universal Robot (UR5 model) is chosen which takes place in the manipulation work. The camera is Intel real-sense D435i, which is used to detect targeted objects. In the end, the completed robot system is as presented in Figure 2.5.

A FRAMEWORK FOR CHANGING WORK FLOORS

To help readers have an overview of how to integrate various single robotic modules to serve a list of designed tasks in the real world without human interventions, a comprehensive system integration for making the mobile

FIGURE 2.5   A mobile manipulator in modern multi-floor buildings.
(Photograph by Nguyen Van Toan.)

manipulator work in modern multi-floor buildings is considered. Namely, an expandable framework is introduced to adapt various tasks as customers' requests. However, the number of modules of a robot system in modern multi-floor buildings is normally large. It is not possible here to present all of them in detail. Without loss of generality, to elucidate the final goal of the book (the robot system integration), the scheme for changing work floors of the mobile manipulator (the elevator operation task) is chosen since it covers almost modules of the proposed robot system. Related works for the elevator operation task can be seen in [1–14]. In [1], an interactively teaching task is introduced to help the mobile robot taking elevators. Thereafter, in [2], a mobile robot without arms is used for the navigation, which paid some attentions for the appearance of elevators in the working environment. In [3], an elevator button-pushing task is presented by using a micro-camera mounted on the end effector of a robot arm. After that, the key challenge of autonomous interaction with an unknown elevator button panel is presented in [4]; a framework including design, implementation, and experimental evaluation of a semi-humanoid robotic system for autonomous multi-floor navigation is presented in [5]; and a wheeled mobile robot for path tracking and for automatically taking an elevator is presented in [6]. To overcome shortcomings of previous studies, a robotic method with navigation and elevator button manipulation abilities for multi-floor building is presented in [7] and is then improved in [8, 9]. Besides, an intelligent mobile robot controller is designed for hotel room service with deep learning arm-based elevator manipulator in [10], a robot button pressing in human environments is presented in [11], and a combination of the faster RCNN architecture and an optimal character recognition for the elevator button recognition is presented in [12]. Most recently, an autonomous eye-in-hand robotic system for elevator button operation is presented in [13], using a deep neural network for elevator button detection, a button pose estimation, and a coarse-to-fine control schema. In [14], a safe human-like elevator button manipulation is presented. After all, a completed framework for the robot system to change its work floors is rarely considered.

Based on our observations, to help the robot system change its working floors successfully, some sets of robotic modules should be considered: (2D1) the mobile robot navigation (self-driving task) to help the robot system travel in its workspace; (2D2) perception works, including elevator button detection, button light-up status recognition, elevator door status checking, human detection, and floor number recognition; (2D3) the manipulator manipulation, such as pressing elevator buttons, or changing the camera view; and (2D4) the integration of (2D1), (2D2), and (2D3) should be conducted to make the robot system work at the high level of the automation. A framework for the mobile manipulator to change its working floors is therefore presented in Figure 2.6, in which the button manipulation process is a combination of works in [12–14]. In brief, the proposed elevator button manipulation procedure in [14] is product-oriented, which meets realistic requirements. However, their proposed elevator button recognition algorithm is easily affected by the glare and the brightness of the environmental light condition. This disadvantage is mainly originated from the mirror-like surface of the elevator button panel. By contrast, the works in [12, 13] have been proved to recognize various types of elevator buttons robustly in various environments. Unfortunately, the button manipulation procedure in these papers is just a research case, and their assumptions are not realistic enough. Therefore, to improve the performance of our proposed robot system, the elevator button recognition in [12, 13] and the elevator button manipulation procedure in [14] are incorporated.

The whole procedure to change work floors is briefly presented as follows: Firstly, the AMR travels to the teach-point in front of the elevator outside panel after it is requested to change the working floor. Since the AMR approached its targeted point, the manipulator adjusts its configuration to set the initial view of the camera so that the elevator button panel appears in the camera frame. Sometimes, the camera cannot see the elevator button panel due to the big orientation error of the AMR. In those cases, the manipulator must implement searching motions to find the location of the elevator button panel. Then, elevator buttons are detected by the camera. And, the position of the targeted button (Up Button or Down Button) is obtained. Next, the

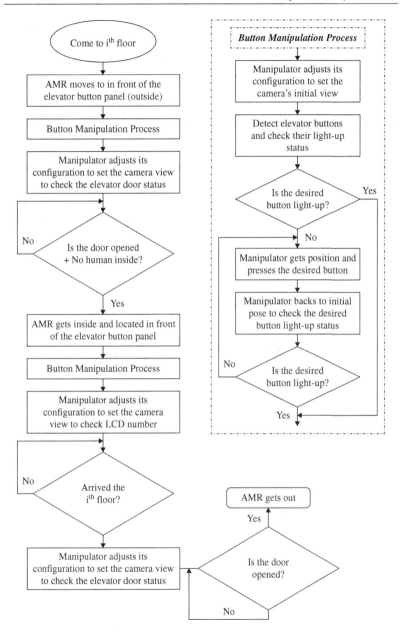

FIGURE 2.6 The scheme of the proposed mobile manipulator to change its working floors

(Photograph by Nguyen Van Toan.)

manipulator presses the targeted elevator button. To ensure that the press action is succeeded, the light-up status of the pressed button is then checked. After the targeted elevator button is pressed successfully, the manipulator adjusts its configuration to change the camera view to check the elevator door status and the appearance of persons before it enters the elevator. Inside the elevator, the mobile manipulator moves in front of the elevator button panel and detects and presses the targeted floor button. Here, the elevator button manipulation process is conducted similarly to what it did outside. Right after the inside targeted button (the targeted floor) is pressed successfully, the manipulator adjusts its configuration to set the camera view to check the floor number, then check the elevator door status and the appearance of persons sequentially before it gets out of the elevator. The robotic software system is conducted on ROS2. Details of software modules (algorithms and settings) and the system integration are presented in the next chapters. The main programs of the presented works can be found in the Appendix section of each chapter. Again, it is reminded that other tasks of the proposed robot system in modern multi-floor buildings can be easily added to the presented expandable framework.

ABBREVIATIONS

2D:	Two Dimensional
3D:	Three Dimensional
LiDAR:	Light Detection and Ranging
AR markers:	ARUCO Markers
RGB-D:	Red-Green-Blue and Depth
AMR:	Autonomous Mobile Robot
FOV:	Field of View
Faster RCNN:	Faster Region-based Convolutional Neural Network
OCR:	Optimal Character Recognition
ROS2:	Robot Operating System 2

REFERENCES

[1] Koji Iwase, Jun Miura, Yoshiaki Shinrai. *Teaching a mobile robot to take elevators*. Mechatronics for Safety, Security and Dependability in a New Era, 2007, pp. 229–234.

[2] Jeong-Gwan Kang, Su-Yong An, Se-Yeong Oh. *Navigation strategy for the service robot in the elevator environment.* In 2007 International Conference on Control, Automation and Systems, Seoul, South Korea, 17–20 October 2007, pp. 1092–1097.

[3] Wen-June Wang, Cheng-Hao Huang, I-Hsian Lai, Han-Chun Chen. *A robot arm for pushing elevator buttons.* In Proceedings of SCIE Annual Conference 2010, Taipei, Taiwan, 18–21 August 2010, pp. 1844–1848.

[4] Ellen Klingbeil, Carpenter Blake, Russakovsky Olga, Ng Andrew. *Autonomous operation of novel elevators for robot navigation.* In IEEE International Conference on Robotics and Automation (ICRA), Anchorage, AK, USA, 03–07 May 2010, pp. 751–758.

[5] Daniel Troniak, Junaed Sattar, Ankur Gupta, James J. Little, Wesley Chan, Ergun Calisgan, Elizabeth Crogt, Machiel Van der Loos. *Charlie rides the elevator-integrating vision, navigation and manipulation towards multi-floor robot locomotion.* In 2013 International Conference on Computer and Robot Vision, Regina, SK, Canada, 28–31 May 2013, pp. 1–8.

[6] Jih-Gau Juang, Chia-Lung Yu, Chih-Min Lin, Rong-Guan Yeh, Imre J. Rudas. *Real-time image recognition and path tracking of a wheeled mobile robot for taking an elevator.* Acta Polytechnica Hungarica, vol. 10, no. 6, pp. 5–23, 2013.

[7] Ali A. Abdulla, Hui Liu, Norbert Stoll, Kerstin Thurow. *A new robust method for mobile robot multifloor navigation in distributed life science laboratories.* Journal of Control Science and Engineering, vol. 2016, Articles ID 3589395, pp. 1–17, 2016.

[8] Ali A. Abdulla, Mohammed M. Ali, Norbert Stoll, Kerstin Thurow. *Integration of navigation, vision, and arm manipulation towards elevator operation for laboratory transportation system using mobile robots.* Journal of Automation, Mobile Robotics & Intelligent Systems, vol. 11, no. 4, pp. 34–50, 2017.

[9] Kerstin Thurow, Lei Zhang, Hui Liu, Steffen Junginger, Norbert Stoll, Jiahao Huang. *Multi-floor laboratory transportation technologies based on intelligent mobile robots.* Transportation Safety and Environment, vol. 1, no. 1, pp. 37–53, 2019.

[10] Po-Yu Yang, Tzu-Hsuan Chang, Yu-Hao Chang, Bing-Fei Wu. *Intelligent mobile robot controller design for hotel room service with deep learning arm-based elevator manipulator.* In 2018 International Conference on System Science and Engineering (ICSSE), New Taipei, Taiwan, 28–30 June 2018, pp. 1–6.

[11] Fan Wang, Gerry Chen, Kris Hauser. *Robot button pressing in human environments.* In 2018 IEEE International Conference on Robotics and Automation (ICRA), Brisbane, QLD, Australia, 21–25 May 2018, pp. 7173–7180.

[12] Delong Zhu, Tingguang Li, Danny Ho, Tong Zhou, Max Q.-H. Meng. *A novel OCR-RCNN for elevator button recognition.* In 2018 IEEE/RSJ International Conference on Intelligent Robots and Systems (IROS), Madrid, Spain, 01–05 October 2018, pp. 3626–3631.

[13] Delong Zhu, Zhe Min, Tong Zhou, Tingguang Li, Max Q.-H. Meng. *An autonomous eye-in-hand robotic system for elevator button operation based on deep recognition network.* IEEE Transactions on Instrumentation and Measurement, vol. 70, Art no. 2504113, pp. 1–13, 2021.

[14] Nguyen Van Toan, Jeong Jin-Hyeon, Jaewon Jo. *An efficient approach for the elevator button manipulation using the visual-based self-driving mobile manipulator.* Industrial Robot: The International Journal of Robotics Research and Application, vol. 50, no. 1, pp. 84-93, 2022. https://doi.org/10.1108/IR-03-2022-0063

Robot System Analysis

3

Phan Bui Khoi

In Chapter 2, a robot system is proposed to work in modern multifloor buildings, which consists of an autonomous mobile robot (the differential drive model) and an eye-on-hand manipulator (6-degrees of freedom [DOF]). Besides, a robotic framework for the proposed mobile manipulator to change its work floors is also presented. In this chapter, the kinematics and dynamics of the proposed robot system are analyzed.

MOBILE ROBOT

The mobile base is firstly considered since the self-driving ability of the mobile manipulator is inevitable to work in multifloor buildings. As mentioned, this autonomous mobile robot is a differential drive model.

Kinematics of Mobile Robot

The coordinate systems and kinematic parameters of the mobile robot are defined in Figure 3.1. Here, the $X_0 Y_0 Z_0$ is the global coordinate system. The $X_C Y_C Z_C$ is the base frame of the mobile robot which is attached to the robot body. The origin of $X_C Y_C Z_C$ is at the point C, which is the midpoint of the axis connecting the centers of two driven wheels of the mobile robot. The specified coordinates of $X_C Y_C Z_C$ with respect to $X_0 Y_0 Z_0$ are defined as x, y, and φ. Rotation angles of the left and right wheels are θ_L and θ_R, respectively. The length of the axle connecting the centers of two driven wheels is denoted as b and the radius of the driven wheel is denoted as r.

DOI: 10.1201/9781003352426-3

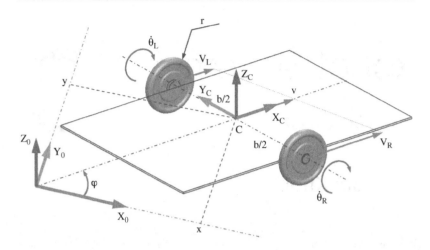

FIGURE 3.1   Coordinates and kinematic parameters of the mobile robot.
(Photograph by Phan Bui Khoi.)

The generalized coordinate vector of the mobile robot can be presented as:

$$q = [x, y, \varphi, \theta_L, \theta_R]^T \tag{3.1}$$

Based on Equation (3.1), the vector of the velocity is denoted as:

$$\dot{q} = [\dot{x}, \dot{y}, \dot{\varphi}, \dot{\theta}_1, \dot{\theta}_R]^T \tag{3.2}$$

It is assumed that (3A1) the robot arm is stationary with respect to the mobile robot while the mobile robot is moving (therefore, the mobile robot and the robot arm are considered as a solid body), and (3A2) there is no slip between the driven wheels and the floor. As a result, the kinematic constraints of the mobile robot can be presented as

$$\begin{bmatrix} \dot{x} \\ \dot{y} \\ \dot{\varphi} \end{bmatrix} = \begin{bmatrix} \dfrac{r}{2}\cos\varphi & \dfrac{r}{2}\cos\varphi \\ \dfrac{r}{2}\sin\varphi & \dfrac{r}{2}\sin\varphi \\ \dfrac{r}{b} & -\dfrac{r}{b} \end{bmatrix} \begin{bmatrix} \dot{\theta}_L \\ \dot{\theta}_R \end{bmatrix} \tag{3.3}$$

It is clear that the linear and angular velocities of the mobile robot body in the left side of Equation (3.3) are obtained if the angular velocities of the driven wheels are given. By contrast, to control the mobile robot, it is required to calculate the motion of the driven wheels through the planned path.

The mobile robot has two degrees of freedom, so its motions can be requested with two independent defined parameters. In general, the robot is supposed to follow a path (curve) with the radius $\rho(t)$, and at a speed v (the center C tends to be on the path). Then, the velocities of two driven wheels can be obtained by:

$$\begin{bmatrix} \dfrac{r}{2} & \dfrac{r}{2} \\ \dfrac{r}{b} & -\dfrac{r}{b} \end{bmatrix} \begin{bmatrix} \dot{\theta}_L \\ \dot{\theta}_R \end{bmatrix} = \begin{bmatrix} 1 \\ \dfrac{1}{\rho} \end{bmatrix} v \tag{3.4}$$

In the case that the angular velocity of the mobile robot is approximately zero (the mobile robot follows a straight line, then the radius $\rho(t) = \infty$), the angular velocity of two driven wheels should be:

$$\dot{\theta}_R = \dot{\theta}_L = \frac{v}{r} \tag{3.5}$$

In the case that the linear velocity of the mobile robot is approximately zero (the center C of the robot is stationary and the robot rotates around the vertical axis Z_C to change the direction with the angular velocity ω), then angular velocities of two driven wheels have opposite signs, same absolute values, and are determined as follows:

$$\dot{\theta}_L = \dot{\theta}_R = \frac{\omega b}{2r} \tag{3.6}$$

Dynamics of Mobile Robot

In general, the mobile robot is a non-holonomic system when it is following a path (a straight line or a curve). Therefore, the Lagrange multiplier can be used to establish the system of differential equations of motion of this non-holonomic system [1–3], as follows:

$$\begin{cases} M(q)\ddot{q} + \psi(q,\dot{q}) + G(q) + Q(q) + Q^c(q) = U \\ J(\varphi)\dot{q} = 0 \end{cases} \tag{3.7}$$

Here, q is the generalized coordinate vector, as presented in Equation (3.1). $M(q)$ is the mass matrix of size (5×5), whose elements are determined as follow:

$$\left\{ \begin{aligned} \mathbf{M}(\mathbf{q})_{(5\times5)} &= \sum_{i=1}^{3} \left(\mathbf{J}_{Ti}^{T} m_i \mathbf{J}_{Ti} + \mathbf{J}_{Ri}^{T} \, {}^{ci}\Theta_{ci} \mathbf{J}_{Ri} \right) \\ \mathbf{J}_{Ti} &= \frac{\partial v_{ci}}{\partial \dot{q}} \\ \mathbf{J}_{Ri} &= \frac{\partial \, {}^{ci}\omega_{ci}}{\partial \dot{q}} \end{aligned} \right. \tag{3.8}$$

In Equation (3.8), C_i is the center of mass of the link i (here, $i = 1, 2, 3$ represent the robot body, the left wheel, and the right wheel, respectively). v_{ci} is the velocity vector of the C_i with respect to the coordinate system $X_0Y_0Z_0$. ${}^{ci}\omega_{ci}$ is the angular velocity of the link i, expressed in the coordinate system of the link i. m_i is the mass of the link i. ${}^{ci}\Theta_{ci}$ is the inertia tensor of the link i with respect to the center of mass C_i, expressed in the ith coordinate system with the origin at the center of mass of the link i.

In Equation (3.7), $\psi(q, \dot{q})$ is the vector of size (5×1), whose elements are generalized forces of Coriolis and centrifugal inertial forces. They are calculated via the elements of matrix $M(q)$:

$$\left\{ \begin{aligned} \psi_j &= \sum_{k,l=1}^{5} (k, l; j) \dot{q}_k \dot{q}_l; \\ (k, l; j) &= \frac{1}{2} \left(\frac{\partial m_{kj}}{\partial q_l} + \frac{\partial m_{lj}}{\partial q_k} - \frac{\partial m_{kl}}{\partial q_j} \right) \end{aligned} \right. \tag{3.9}$$

Here, $(k, l; j)$ is the Christoffel notation with three indexes of the first kind. m_{kl} $(k, l = 1, \ldots, 5)$ are elements of the matrix $M(q)$.

It is assumed that the mobile robot works on a horizontal floor, and there are no elastic forces. In addition, other nonconservative forces are ignored. Therefore, the components $G(q)$ and $Q(q)$ in Equation (3.7) are zero. Besides, the motion effects of four casters are also ignored since they are very small in comparison with the robot body. Then, their masses are included in the mass of the robot body.

U is the vector (5×1) of the generalized forces of the driven forces/torques, in which only the driven moments τ_L and τ_R of the motors (relative to the coordinates θ_L, θ_R) are nonzero components.

$$U = \left[0, 0, 0, \tau_L, \tau_R \right]^T \tag{3.10}$$

The generalized coordinates in Equation (3.1) are not independent, and their relationship is presented by three non-holonomic constraints in Equation (3.3). Thus, $Q_c(q)$ is used as the generalized force vector representing the effect of the non-holonomic constraints in Equation (3.3).

Another form of the system of Equation (3.3) can be presented as:

$$
\begin{bmatrix}
1 & 0 & 0 & -\dfrac{r}{2}\cos\varphi & -\dfrac{r}{2}\cos\varphi \\
0 & 1 & 0 & -\dfrac{r}{2}\sin\varphi & -\dfrac{r}{2}\sin\varphi \\
0 & 0 & 1 & -\dfrac{r}{b} & \dfrac{r}{b}
\end{bmatrix}
\begin{bmatrix}
\dot{x} \\
\dot{y} \\
\dot{\varphi} \\
\dot{\theta}_L \\
\dot{\theta}_R
\end{bmatrix} = 0 \tag{3.11}
$$

Now, $J(\varphi)$ is used to present the first element of Equation (3.11). It is also the first element of the second line of Equation (3.7):

$$
J(\varphi) =
\begin{bmatrix}
1 & 0 & 0 & -\dfrac{r}{2}\cos\varphi & -\dfrac{r}{2}\cos\varphi \\
0 & 1 & 0 & -\dfrac{r}{2}\sin\varphi & -\dfrac{r}{2}\sin\varphi \\
0 & 0 & 1 & -\dfrac{r}{b} & \dfrac{r}{b}
\end{bmatrix}, \tag{3.12}
$$

The defined expression $Q_c(q)$ can be obtained as follow:

$$
Q_c(q) = J^T(\varphi)\lambda \tag{3.13}
$$

Here J^T is the transpose of J and $\lambda(3 \times 1)$ is the vector of Lagrange multipliers.

Based on Equations (3.7), (3.11), (3.12), and (3.13), the system of dynamical equations of the mobile robot is obtained as:

$$
\begin{cases}
M(q)\ddot{q} + \psi(q,\dot{q}) + J^T(\varphi)\lambda = U \\
J(\varphi)\dot{q} = 0
\end{cases} \tag{3.14}
$$

The solution of the system of Equation (3.14) is quite complicated, which causes the control solution to be complicated as well. Practically, to simplify the calculation, the control method can be applied with a division of the curvilinear motion of the mobile robot into separated rotations and translations. By doing this, the robot system becomes a holonomic system, and the differential equations of motion become more simply. This simplification is consistent with the mobile robot's ability to locate by two degrees of freedom in the real working environment. A motion strategy can be presented as follows: (step 1) the robot rotates to determine the moving direction, (step 2) the robot moves to the targeted position, and (step 3) the robot rotates to correct the designed direction. For steps 1 and 3, the angular velocities of two driven wheels have opposite directions and their values are the same. The motion equation of the robot is now presented as follows:

$$\left(J_{bz} \frac{2r^2}{b^2} + \frac{5mr^2}{8} + \frac{mr^4}{2b^2} \right) \ddot{\theta} = \tau \tag{3.15}$$

Here, J_{bz} is the mass moment of inertia of the robot body (including the mass of the manipulator) about the axis Z_c. m is the mass of each wheel. $\ddot{\theta}$ is the algebraic angular acceleration of the driven wheel, which is positive or negative depending on the direction of rotation of the robot. τ is the vector of torques of driven motors (for driven wheels). It is noted that driven torques and rotation directions of two wheels have the same magnitude and opposite directions.

For the translation motion in step 2, the differential equation of motion of the robot is

$$\frac{(M+3m)r^2}{2} \ddot{\theta} = \tau \tag{3.16}$$

Here, M is the mass of the robot body. In translation motion, two driven wheels have the same angular velocity and driven torque (both their directions and values).

MANIPULATOR

It is assumed that the mobile robot is stationary when the manipulator is working, and vice versa. Therefore, the mobile robot can be considered as the fixed base of the manipulator.

Kinematics of Manipulator

The structure of the robot arm and its coordinate systems are presented in Figure 3.2. Here, the base coordinate system $X_0Y_0Z_0$ is attached to the fixed

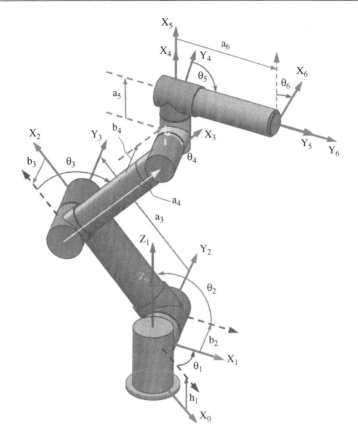

FIGURE 3.2 The kinematic structure of the manipulator.

(Photograph by Phan Bui Khoi.)

link. Its origin is at the specified position on the body of the mobile robot. The coordinate systems $X_iY_iZ_i$ are attached to the link i, $i = 1, ..., 6$, respectively, called link coordinate systems. Since the third axis of the coordinate system is determined by the right-hand rule (depending on two remaining axes), only two axes are shown in the figure.

The relative position and orientation of the ith coordinate system in the $(i-1)$th coordinate system are presented as follows: (3B1) angles of roll, pitch, and yaw are used for the relative orientation, and (3B2) coordinates of the origin of the ith coordinate system with respect to the $(i-1)$th coordinate system are used for the relative position. The kinematic relationship of all coordinate systems of the manipulator can be seen in Table 3.1, where each

TABLE 3.1   The kinematic relationship of the manipulator

	ROLL	PITCH	YAW	X	Y	Z
shoulder	0	0	θ_1	0	0	h_1
upper_arm	0	θ_2	0	0	b_2	0
forearm	0	θ_3	0	a_3	b_3	0
wrist_1	0	θ_4	0	a_4	b_4	0
wrist_2	θ_5	0	0	a_5	0	0
wrist_3	0	θ_6	0	0	b_6	0

row represents the state between two consecutive coordinate systems. For example, the third row represents rotation and translation from the coordinate systems $X_2Y_2Z_2$ to the coordinate systems $X_3Y_3Z_3$ according to the angles of roll, pitch, and jaw and x, y, z coordinates, as denoted by

$$^2p_3 = \left[0, \theta_3, 0, a_3, b_3, 0\right]^{\mathrm{T}} \tag{3.17}$$

The vector $(i\text{-}1)p_i$ has six elements representing the rotations and translations of the $(i\text{-}1)$th coordinate system to the ith coordinate system. In fact, depending on the structure of links and joints between two coordinate systems, there are usually less than six coordinate transformations. Same as the vector $^2p_3$, it only represents a rotation of an angle θ_3 around the Y_2 axis, and translations of lengths a_3, b_3, along the x_2 and y_2 axes, respectively.

In this section, dual quaternion algebra is used to investigate the kinematics of the manipulator [4–9]. The coordinate transformation from the $(i\text{-}1)$th system to the ith system can be represented by dual quaternion $^{i-1}q_i$ as follows:

$$^{i-1}q_i = {}^{i-1}r_i + \epsilon \frac{1}{2} {}^{i-1}t_i^{i-1}\,{}^{i-1}r_i \tag{3.18}$$

$$^{i-1}r_i = \left(\cos\frac{\theta_i}{2}, u_i \sin\frac{\theta_i}{2}\right), \tag{3.19}$$

where ϵ is the dual unit, $\epsilon^2 = 0$, $\epsilon \neq 0$.

The matrix form of Equation (3.19) is presented as follows:

$$^{i-1}r_i = \left[\cos\frac{\theta_i}{2}, u_{ix} \sin\frac{\theta_i}{2}, u_{iy} \sin\frac{\theta_i}{2}, u_{iz} \sin\frac{\theta_i}{2}\right]^{\mathrm{T}} \tag{3.20}$$

Here, $^{i-1}r_i$ is a quaternion, representing the rotation of the $(i\text{-}1)$th coordinate system at an angle $\theta_i/2$, around the axis of rotation defined by the unit vector u_i. For example, the unit vector representing the rotation of the

TABLE 3.2   The dual quaternion algebra presentation of the manipulator

	$\theta_i/2$	U_{iX}	U_{iY}	U_{iZ}	X	Y	Z
			U_i			$^{i-1}t_i^{i-1}$	
shoulder	θ_1	0	0	1	0	0	h_1
upper_arm	θ_2	0	1	0	0	b_2	0
forearm	θ_3	0	1	0	a_3	b_3	0
wrist_1	θ_4	0	1	0	a_4	b_4	0
wrist_2	θ_5	1	0	0	a_5	0	0
wrist_3	θ_6	0	1	0	0	b_6	0

coordinate system $X_0Y_0Z_0$ by an angle $\theta_1/2$ about the axis Z_0 is defined as follows:

$$u_1 = 0.u_{ix} + 0.u_{iy} + 1.u_{iz} \tag{3.21}$$

$^{i-1}t_i^{i-1}$ is a position quaternion, which represents the position of the origin of the system $X_iY_iZ_i$ with respect to the system $X_{i-1}Y_{i-1}Z_{i-1}$. This position quaternion is expressed in the coordinate system $X_{i-1}Y_{i-1}Z_{i-1}$. Equation (3.18) implies that the frame rotation is carried out first, and then a translation is carried out relative to the reference frame.

The components of the coordinate transformations (in Equation (3.18)) of the robot manipulator are presented in Table 3.2. And, the composition of the coordinate transformations from the $X_0Y_0Z_0$ coordinate system to the $X_6Y_6Z_6$ coordinate system is presented in Equation (3.22), which determines the pose of the end effector with respect to the base coordinate system:

$$^0q_6 = {}^0q_1 \, {}^1q_2 ... {}^5q_6 \tag{3.22}$$

From Equation (3.22), the position and orientation of the end-effector are determined by the composite dual quaternion, which is represented as follows:

$$^0q_6 = {}^0r_6 + \epsilon \frac{1}{2} {}^0t_6^0 \, {}^0r_6 \tag{3.23}$$

Here, $^0r_6$ is the quaternion representing the orientation of the coordinate system $X_6Y_6Z_6$ of the end effector with respect to the base coordinate system $X_0Y_0Z_0$. $^0t_6^0$ is a vector quaternion representing the origin of the coordinate system $X_6Y_6Z_6$ with respect to the system $X_0Y_0Z_0$, which is expressed in the system $X_0Y_0Z_0$.

Equation (3.23) can be used to solve the robot's forward kinematics. By contrast, if the robot's manipulation target is defined by a dual quaternion $^0p_6$, Equation (3.23) represented in form of Equation (3.24) allows to solve the robot's inverse kinematics.

$$^0r_6 + \epsilon \frac{1}{2}\,{}^0t_6^0\,{}^0r_6 = {}^0p_6 \tag{3.24}$$

The angular and linear velocities of the end effector are determined by the time derivative of the dual quaternion:

$$^0\dot{q}_6 = {}^0\dot{r}_6 + \epsilon \frac{1}{2}\left({}^0\dot{t}_6^0\,{}^0r_6 + {}^0t_6^0\,{}^0\dot{r}_6\right) = \frac{1}{2}\,{}^0\Upsilon_6^0\,{}^0q_6 \tag{3.25}$$

where

$$^0\dot{r}_6 = \frac{1}{2}\,{}^0\omega_6^0\,{}^0r_6 \tag{3.26}$$

$$^0\Upsilon_6^0 = {}^0\omega_6^0 + \epsilon\left({}^0v_6^0 + {}^0t_6^0 \times {}^0\omega_6^0\right) \tag{3.27}$$

$^0\omega_6^0, {}^0v_6^0$ are the angular velocity and the linear velocity of the end effector with respect to the base coordinate system $X_0Y_0Z_0$, respectively, expressed in the base coordinate system $X_0Y_0Z_0$.

To solve the dynamic problem, it is necessary to determine the angular and linear velocities of the links. To do so, each manipulator's link is attached by a coordinate system with the origin at the center of mass of the link, and its axes are parallel to the axes of the link coordinate system. Such a coordinate system is called the center of mass coordinate system of the link.

Figure 3.3 shows an example of the link coordinate system, the center of mass coordinate system, and the parameters representing the center of mass coordinate system with respect to the link coordinate system. The dual quaternion defines the coordinate system of the center of mass of the ith link with respect to the base coordinate system:

$$^0q_{Ci} = {}^0q_i\,{}^iq_{Ci} \tag{3.28}$$

Here, $^iq_{ci}$ is a dual quaternion representing the center of the mass coordinate system with respect to the link coordinate system.

The link's angular velocity and linear velocity of the center of mass of links can be obtained from the time derivative of the dual quaternion in Equation (3.28).

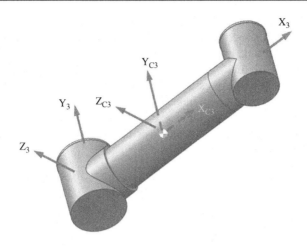

FIGURE 3.3   The definition and the relationship of the link coordinate system and the center of mass coordinate system.

(Photograph by Phan Bui Khoi.)

Dynamics of Manipulator

The manipulator is a serial structure (open chain), so the Lagrange differential equation of motion of the second kind is convenient to use for the dynamic problem [10–20].

$$M(q)\ddot{q} + \psi(q,\dot{q}) + G(q) + Q(q) = U \tag{3.29}$$

where q is the generalized coordinate vector, as presented:

$$q = \left[q_1, q_2, \cdots, q_6\right]^T = \left[\theta_1, \theta_2, \cdots, \theta_6\right]^T \tag{3.30}$$

$M(q)$ is the mass matrix of size (6×6), whose elements are determined as follow:

$$
\begin{cases}
M(q)_{(6\times6)} = \sum_{i=1}^{6} \left(J_{Ti}^T m_i J_{Ti} + J_{Ri}^T {}^{ci}\Theta_{ci} J_{Ri}\right) \\[2mm]
J_{Ti} = \dfrac{\partial v_{ci}}{\partial \dot{q}} \\[2mm]
J_{Ri} = \dfrac{\partial {}^{ci}\omega_{ci}}{\partial \dot{q}}
\end{cases}
\tag{3.31}
$$

where C_i, $i = 1, ..., 6$ is the center of mass of the ith link. v_{ci} is the vector of velocity of C_i with respect to the coordinate system $X_0 Y_0 Z_0$. $^{ci}\omega_{ci}$ is the angular velocity of the ith link, expressed in the coordinate system of the ith link. m_i is the mass of the ith link. $^{ci}\Theta_{ci}$ is the inertia tensor of the ith link with respect to the center of mass C_i, expressed in the ith coordinate system with the origin at the center of mass of the ith link.

$\psi(q, \dot{q})$ (6×1) is the vector of size (6×1), whose elements are generalized forces of Coriolis and centrifugal inertial forces. Their values are calculated via the elements of matrix $M(q)$:

$$\begin{cases} \psi_j = \sum_{k,l=1}^{6} (k,l;j) \dot{q}_k \dot{q}_l; \\ (k,l;j) = \frac{1}{2}\left(\frac{\partial m_{kj}}{\partial q_l} + \frac{\partial m_{lj}}{\partial q_k} - \frac{\partial m_{kl}}{\partial q_j} \right) \end{cases} \tag{3.32}$$

where $(k, l; j)$ is the Christoffel notation with three indexes of the first kind. m_{kl} $(k, l = 1, ..., 6)$ are the elements of the matrix $M(q)$.

$G(q)$ (6×1) is the vector of the generalized forces of the conservative forces acting on the robot.

$$G(q)_{(6\times1)} = [G_1, G_2, .., G_6]^T; \quad G_j = \frac{\partial \Pi}{\partial q_j} \tag{3.33}$$

$Q(q)$ (6×1) is the vector of the generalized forces of the nonconservative forces acting on the robot. Here, frictional forces and resistance are ignored. However, the nonconservative force may appear when the end effector acts on elevator buttons. Although this force is quite small, it needs to be included in the dynamic equations because it is related to safety.

$$Q = J_{TE}^T F; \quad J_{TE} = \frac{\partial r_E}{\partial q}; \quad F = [F_x, F_y, F_z]^T \tag{3.34}$$

where J_{TE} is the Jacobi matrix of the coordinate vector of the robot's impact point of the end effector. J_{TE}^T is the transpose of the matrix $\underline{J}_{TE}$. F is the force that the end effector acts on elevator buttons.

U (6×1) is the vector of the generalized forces of the driven forces/torques. Here, the driven force is the motor moment M_i at joints.

$$U = [M_1, M_2, ..., M_6]^T \tag{3.35}$$

ABBREVIATION

DOF: Degrees of Freedom

REFERENCES

[1] Anatolii Isakovich Lurie. *Analytical mechanics: foundations of engineering mechanics.* 2002nd edn., Springer, Cham, 2002.

[2] Carl S. Helrich. *Analytical mechanics.* Springer, Cham, 2016.

[3] Nivaldo A. Lemos. *Analytical mechanics.* Cambridge University Press, Cambridge, 2018.

[4] M. A. Clifford. *Preliminary sketch of biquaternions.* In Proceedings of the London Mathematical Society, 1871, s1-4.1, pp. 381–395. https://doi.org/10.1112/plms/s1-4.1.381

[5] Konstantinos Daniilidis. *Hand-eye calibration using dual quaternions.* The International Journal of Robotics Research, vol. 18, no. 3, pp. 286–298, 1999.

[6] Bedia Akyar. *Dual quaternions in spatial kinematics in an algebraic sense.* Turkish Journal of Mathematics, vol. 32, no. 4, pp. 373–391, 2008.

[7] Mahmoud Gouasmi, Mohammed Ouali, Fernini Brahim. *Robot kinematics using dual quaternions.* International Journal of Robotics and Automation (IJRA), vol. 1, no. 1, pp. 13–30, 2012.

[8] Erol Özgür, Youcef Mezouar. *Kinematic modeling and control of a robot arm using unit dual quaternions.* Robotics and Autonomous Systems, vol. 77, pp. 66–73, 2016. http://dx.doi.org/10.1016/j.robot.2015.12.005

[9] Bruno Vilhena Adorno. *Robot kinematic modeling and control based on dual quaternion algebra—Part I: fundamentals.* HAL 2017. https://hal.archives-ouvertes.fr/hal-01478225

[10] John J. Craig. *Introduction to robotics: mechanics and control.* 3rd edn., Pearson, London, 1989.

[11] Lung-Wen Tsai. *Robot analysis: the mechanics of serial and parallel manipulators.* John Willey & Sons, New York, 1999.

[12] Bruno Siciliano, Loreno Sciavicco, Luigi Villani, Giuseppe Oriolo. *Robotics: modelling, planning and control.* Springer, London, 2009.

[13] Bruno Sicilianno, Oussama Khatib. *Springer handbook of robotics.* Springer-Verlag, Berlin-Heidelberg, 2008.

[14] Mark W. Spong, Seth Hutchinson, Mathukumalli Vidyasagar. *Robot modeling and control.* 1st edn., Wiley, Hoboken, 2020.

[15] Peter Corke. *Robotics and control. Subtitle: Fundamental algorithms in MATLAB.* Springer, Cham, 2022.

[16] Jose M. Pardos-Gotor. *Screw theory in robotics: an illustrated and practicable introduction to modern mechanics.* 1st edn., CRC Press, Boca Raton, 2022.

[17] Jaime Gallardo-Alvarado, José Gallardo-Razo. *Mechanisms: kinematic analysis and applications in robotics.* 1st edn., Academic Press: An imprint of Elsevier, Cambridge, 2022.
[18] Phan Bui Khoi, Nguyen Van Toan. *Hedge-algebras-based controller for mechanisms of relative manipulation.* International Journal of Precision Engineering and Manufacturing, vol. 19, no. 3, pp. 377–385, 2018. doi:10.1007/s12541-018-0045-8.
[19] Nguyen Van Toan, Phan Bui Khoi. *A control solution for closed-form mechanisms of relative manipulation based on fuzzy approach.* International Journal of Advanced Robotic Systems, vol. 16, no. 2, pp. 1–11, 2019. https://doi.org/10.1177/1729881419839810
[20] Phan Bui Khoi, Ha Thanh Hai, Hoang Vinh Sinh. *Eliminating the effect of uncertainties of cutting forces by fuzzy controller for robots in milling process.* Applied Sciences, vol. 10, no. 5, Article number: 1685, 2020. https://doi.org/10.3390/app10051685

Mobile Robot Navigation

4

Nguyen Van Toan

The mobile robot navigation is a crucial capacity of the mobile manipulator to localize its own positions in the reference frame and then plan a collision-free path toward some goals. This chapter presents some core features of the robot navigation, including localization and mapping, global positioning, environmental modeling, path generation, and following with the collision avoidance.

LOCALIZATION AND MAPPING

Localization and mapping are core technologies which help robots to explore and understand their working environments. This task includes many steps, and different steps can be implemented by various algorithms. In general, the localization and mapping can be described by four steps: (step 4A1) collect the data about environment by using sensors, (step 4A2) update the current state by using the odometry data, (step 4A3) update the estimated state from observation landmarks, and (step 4A4) add new observation landmarks into the current state. Nowadays, both visual SLAM and two-dimensional (2D) LiDAR SLAM have received a great attention from developers, researchers, and scientists. First of all, a comparison of modern general-purpose visual SLAM approaches such as ORB-SLAM3, OpenVSLAM, and RTABMap can be found in [1]. These methods have been proven to provide potential applications in robotics. Besides, some other well-known methods can be mentioned such as gmapping in [2, 3], hector slam in [4, 5], cartographer in [6, 7], or slam toolbox in [8, 9].

DOI: 10.1201/9781003352426-4

In details, the main purpose of robot localization is to estimate robot state at the current time-step and give knowledge about all its measurements $Z^k = \{z_k, i = 1...k\}$. As presented in the previous chapter (Robot System Analysis), the state vector of the mobile robot should include its position and orientation, denoted as $s = [x, y, \varphi]^T$. This state estimation can be considered like a Bayesian filtering problem. Here, the posterior density $p(s_k \mid Z^k)$ of the current state is concerned to obtain the current pose of the robot. As this density is multimodal during the global localization phase, the robot localization is therefore obtained by recursively computing the $p(s_k \mid Z^k)$ at each time-step. Firstly, only the robot motion is modeled to predict the current position of the robot in the form of $p(s_k \mid Z^{k-1})$. The previous measurement is defined as $z^{k-1} = (s_{k-1}, u_{k-1})$. Here s_{k-1} is the previous state and u_{k-1} is the control state. If the current state is only dependent on the previous measurement $z^{k-1} = (s_{k-1}, u_{k-1})$, then the density form of the robot motion model is $p(s_k \mid s_{k-1}, u_{k-1})$. The current density over the state s_k is then calculated by

$$p(s_k \mid Z^{k-1}) = \int p(s_k \mid s_{k-1}, u_{k-1}) p(s_{k-1} \mid Z^{k-1}) ds_{k-1} \qquad (4.1)$$

Secondly, the information from the sensor is incorporated to measure the posterior density $p(s_k \mid Z^k)$. As in [10], it is assumed that the current measurement is independent on previous measurement Z^{k-1}, and given in the term of a likelihood $p(z_k \mid s_k)$. It means that z_k was observed when the robot is at s_k. Using the Bayes theorem, the $p(s_k \mid Z^k)$ is measured as

$$p(s_k \mid Z^k) = \frac{p(z_k \mid s_k) p(s_k \mid Z^{k-1})}{p(z_k \mid Z^{k-1})} \qquad (4.2)$$

Equation (4.2) is used to recursively compute the current state of the robot, and the initial state s_0 is assumed to be identified in the form of the $p(s_0)$.

To date, there is a great amount of studies in the mobile robot localization. It is not possible here to describe all of those methods. In this section, the AMCL (Adaptive Monte Carlo Localization) is preferred to present since it has recently been applied with great practical success and is mainly used for the proposed robot system. It is noted that the Monte Carlo Localization (MCL) is a version of the SIR (sampling/important re-sampling) method [11] and also known as a particle filter. In brief, this filter recursively computes importance sampling approximations of the filter distribution π_k of s_k:

$$\pi_k(s_k \mid Z^k) \approx \hat{\pi}_k(s_k \mid Z^k) = \sum_{i=1}^{K} W_k^i \Delta_{S_k^i}(s_k) \qquad (4.3)$$

Here $W_k^i \geq 0$ are random weights and $\sum_{i=1}^{K} W_k^i = 1$, S_k^i are random variables called *particle* and Δ_s is the point mass at s. At the initial step, *particles* are drawn from π_0 (the initial distribution of s_0) and set $W_0^i = \frac{1}{K}$. At the step k, it is started with $\hat{\pi}_{k-1}$ and draw independently new particles S_k^i from $P\left(\bullet \mid S_{k-1}^i\right)$ (the transition kernel for the value S_{k-1}^i). The particle filter is therefore basically implemented through three steps: (step 4B1) draw $\left(S_{k-1}^{*1},\ldots,S_{k-1}^{*K}\right)$ from $\hat{\pi}_{k-1}$, (step 4B2) draw S_k^i from $P\left(\bullet \mid S_{k-1}^{*i}\right)$ independently for different indices i, and (step 4B3) set $W_k^i \propto p\left(z_k \mid S_k^i\right)$. To deal with the unbalance of weights, some alternative methods are also presented such as auxiliary particle filter [12], resample moves [13], ensemble Kalman filter [14–16], particle smoothing [17, 18], and particle Markov chain Monte Carlo [19]. Furthermore, to increase the efficiency of the particle filter, an adaptive approach is presented in [20], named Kullback-Leiber distance sampling. This particle filter adaptively chooses the number of samples over time. Namely, a small number of samples is chosen if the density is focused on a small part of the state space. Otherwise, a large number of samples are chosen. The number of samples is chosen so that the distance between the maximum likelihood estimate based on samples and the true posterior is less than a threshold ε. This distance is measured by the Kullback-Leibler distance. To have a view on the Adaptive Particle Filter, a brief description about it is now presented. At first, it is assumed to draw m samples from a discrete distribution with k different bins. Then, the number of samples drawn from each bin is denoted as $\underline{S} = (S_1 \ldots, S_k)$. The vector of the probability of $\underline{S}$ is denoted as $\underline{p} = p_1 \cdots p_k$; here $p_1 \ldots p_k$ specifies the probability of each bin. And, the function $\hat{\underline{p}} = n^{-1}\underline{S}$ is used to estimate the maximum likelihood of $\underline{p}$. In addition, the likelihood ratio statistic λ_m for testing $\underline{p}$ is

$$\log \lambda_m = \sum_{j=1}^{k} S_j \log\left(\frac{\hat{p}_j}{p_j}\right) = m \sum_{j=1}^{k} \hat{p}_j \log\left(\frac{\hat{p}_j}{p_j}\right) \tag{4.4}$$

This likelihood ratio converges to a chi-square distribution if $\underline{p}$ is the true distribution; it should be

$$2 \log \lambda_m \rightarrow_d S_{k-1}^2 \quad \text{as} \quad m \rightarrow \infty \tag{4.5}$$

The Kullback-Leibler distance between the maximum likelihood estimate and the true distribution p is represented in Equation (4.4), which is $KL(\hat{p}, p) = \sum_{j=1}^{k} \hat{p}_j \log\left(\frac{\hat{p}_j}{p_j}\right)$. For m, samples are drawn from the true distribution; the probability with the Kullback-Leibler distance smaller than threshold ε is measured via Equations (4.4) and (4.5), as

$$P_{\underline{p}}(KL(\hat{\underline{p}}, \underline{p}) \leq \varepsilon) = P_{\underline{p}}\left(2mKL(\hat{\underline{p}}, \underline{p}) \leq 2m\varepsilon\right) \doteq P\left(S_{k-1}^2 \leq 2m\varepsilon\right) \tag{4.6}$$

Then, the quantiles of the chi-square distribution are

$$P\left(S_{k-1}^2 \le S_{k-1,1-\delta}^2\right) = 1-\delta \tag{4.7}$$

If m samples are chosen so that $2m\varepsilon = S_{k-1,1-\delta}^2$, then the probability with the Kullback-Leibler distance smaller than threshold ε is now:

$$P_p\left(KL(\hat{\underline{p}},\underline{p}) \le \varepsilon\right) \doteq 1-\delta \tag{4.8}$$

The chosen number of samples m as above guarantees the Kullback-Leibler distance smaller than threshold ε, whose value is approximated by the formula in [21], as

$$m = \frac{1}{2\varepsilon}S_{k-1,1-\delta}^2 \doteq \frac{k-1}{2\varepsilon}\left\{1 - \frac{2}{9(k-1)} + \sqrt{\frac{2}{9(k-1)}}z_{1-\delta}\right\}^3 \tag{4.9}$$

Here, $z_{1-\delta}$ is the upper $1-\delta$ quantile of the standard normal $N(0,1)$ distribution. The number k of bins is estimated by counting the number of bins with support during sampling. By doing this, the true posterior distribution is unnecessary to be known. Namely, the number k is estimated through update steps of the particle filter, with the proposal distribution $p\left(s_k \mid s_{k-1},u_{k-1}\right)Bel\left(s_{k-1}\right)$, in which $Bel(\bullet)$ is the instance of the belief.

The adaptive approach in [20] is then used for MCL, called AMCL [22]. The AMCL has recently been applied with great practical success since it is computationally efficient and represents almost arbitrary distributions. Here, the belief $Bel(\bullet)$ is represented by a set of K weighted, random samples, or particles $S = \left\{S_k^i \mid i = 1...K\right\}$, as presented in Equation (4.3). Therefore, samples in MCL include the robot state $s = [x,y,\varphi]^T$ and weights W_k^i. As presented above, the robot localization is considered in two phases: prediction phase and update phase, in which the prediction phase is conducted by using a motion model (related to the robot motion) and the update phase is conducted by using a measurement model (related to the sensor reading). In the prediction phase, the MCL generates K new samples when the robot is moving. This process depends on the previous robot state s_{k-1} and the previous control state u_{k-1}. Samples are generated as similar to (step 4B1) and (step 4B2) of the mentioned SIR particle filter. In the update phase, the sensor measurement is used to re-weighting the sample set, as in (step 4B3) of the mentioned SIR particle filter, where z_k is the sensor measurement and S_k^i is the new sample. It is noted that a normalization constant should be used to enforce the condition $\sum_{i=1}^{K} W_k^i = 1$. For the efficiency of the update phase, the proposed method in [23] can be a good approach. In addition, to relocalize the robot in some cases that the robot loses tracks of its position, a small number of uniformly distributed random samples are added after each estimation step to avoid zero samples. Furthermore, the adaptive factor of the AMCL is created by using the presented approach in [20], which is briefly performed in Equations (4.4)–(4.9).

Next, the robotic mapping task should be concerned. The purpose of this work is to construct a map of the robot-working environment which will be used for motion planning and global localization. It is noted that only indoor mapping is considered since the working environment of the proposed robot system is in modern multifloor buildings. Through the design of the proposed mobile manipulator and its working environment (wide range), 2D LiDAR-based SLAM methods are most suitable for this robot system. As mentioned, the mapping feature of some well-known methods such as gmapping in [2, 3], hector slam in [4, 5], cartographer in [6, 7], or slam toolbox in [8, 9] can be considered because their 2D LiDAR approaches have been popularly used in real robotic applications, both in industrial and social services with great practical success. Grid mapping uses the particle filter and works well in planar environments. Hector slam combines a robust scan-matching approach using a LiDAR system with a three-dimensional (3D) attitude estimation system based on inertial sensing. By contrast, in cartographer, the particle filter is not employed, whose purpose is to adapt modest hardware systems. Since the scan matcher only depends on recent scans, the accumulated error of the pose estimation in the global frame is overcome by a proposed pose optimization, as in [6]. An experimental comparison regarding the performance (including accuracy and time complexity) of gmapping, hector slam, cartographer, and slam toolbox is conducted by using the proposed mobile manipulator. In the result, the cartographer is found to provide better performance for our robot system and its working environments (inside multi-floor buildings or factories with a wide range), in comparison with remaining methods. One of the reasons for the comparison results is due to our robot system's modest hardware. Regarding the initial assessment, the proposed localization method in cartographer [6] seems better than the AMCL to use for the proposed robot system. However, the localization method in [6] seems not long-life because its accumulated error is quite big after the robot has worked for some hours in wide-range environment. Therefore, the cartographer is only used to construct the working environment of the proposed robot system. Then, this constructed map and the AMCL will be used for the robot navigation.

To make the mapping process more convenient, a user interface is created by using Qt5 and linked to cartographer core as shown in Figure 4.1, where the robot namespace and desired actions can be conducted by using defined buttons. This dialog is also used to switch robot working modes: NAV (navigation mode), SLAM (mapping mode), and Restart (restart the robot).

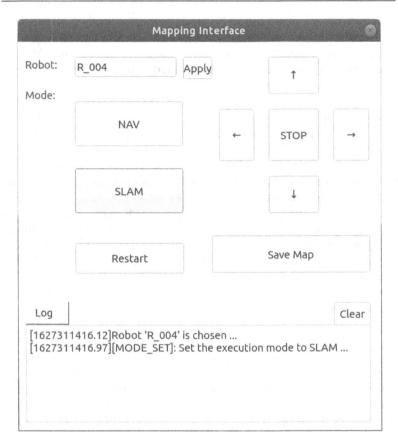

FIGURE 4.1 Mapping user interface.

(Photograph by Nguyen Van Toan.)

Steps should be conducted to start the mapping work: (step 4C1) input the namespace of the used robot into "Robot" text box (in Figure 4.1, the robot R_004 is chosen), (step 4C2) click the "Apply" button, (step 4C3) click the "SLAM" button, (step 4C4) move the robot around its environments by using arrows "Up," "Down," "Left," "Right," and "Stop" (publish linear/angular velocities by using "*cmd_vel*" topic), and (step 4C5) click "Save Map" button after the mapping work is finished to save the constructed map to the desired directory to use for next works. It is reminded that the cartographer package must be firstly installed using [7]. The communication between the user interface in Figure 4.1 and the cartographer core can be conducted in Appendix 4.1, where *RESTART_DIR* is the path to "*robot_start.launch.xml*" which is used to activate the robot system, including LiDAR sensors, cameras, motor control module, robot model description, robot state, and odometry publisher (call-related packages to activate robot devices). *NAV_DIR* is the path to "*nav.launch.xml*" which is used to activate navigation mode for the robot, linked to Navigation2 packages. And, *SLAM_DIR* is the path to "*slam_2d. launch.xml*," linked to the cartographer core to activate the mapping work. For the mapping work, the content of "*slam_2d.launch.xml*" can be seen in Appendix 4.2. The content of "*slam_2d.lua*" in Appendix 4.2 is presented in Appendix 4.3, where "*map_builder.lua*" and "*trajectory_builder.lua*" can be found in the folder "*configuration_files*" of the cartographer package. The SLAM parameters should be matched with the robot hardware and its working environment to ensure the desired accuracy and computational cost. For our system, the LiDAR sensor is TIM561 (Sick), the computer is Intel NUC 8th core i7, and the robot usually works in wide space. Therefore, the maximum range is 25 m, the minimum range is 0.1 m, and the resolution is 0.02. Other parameters can be seen in Appendices 4.2 and 4.3.

Figure 4.2 is an example of the mapping work using cartographer method. After the mapping is finished, this map should be saved in the desired folder to use for navigation work. To do this, click the "Save Map" button in Figure 4.1, which conducts the save_map function in Appendix 4.4 (where the converter from ros2 message to json can be seen in [24]). The file *map.jpg* is saved in the *MAP_PATH*, which is later loaded to map server via "*map.yaml*" file:

```
map_server:
  ros__parameters:
    use_sim_time: True
    yaml_filename: "map.yaml"
    save_map_timeout: 5.0
```

And, the content of "*map.yaml*" is presented as

```
image: /MAP_PATH/map/map.jpg
resolution: 0.050000
origin: [-59.490106, -120.177722, 0.000000]
negate: 0
occupied_thresh: 0.65
free_thresh: 0.196
```

During the navigation period, the AMCL is used for the localization, whose purpose is to estimate the position and orientation of the robot in real time relative to an external reference frame, given a map of the environment and sensor data. To serve navigation works, Navigation2 in [25, 26] is leveraged. The AMCL is conducted by using "*nav2_amcl*" package. A set

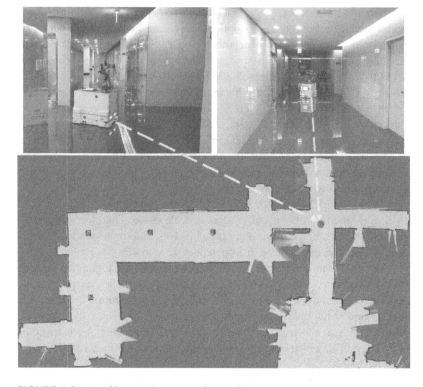

FIGURE 4.2   Working-environmental mapping.

(Photograph by Nguyen Van Toan.)

of AMCL parameters is called from "*nav2_params.yaml*," as presented in Appendix 4.5. These parameters should also be tuned based on the hardware system of the robot.

ENVIRONMENTAL MODELING

The robot model, robot states, and environmental information should be updated frequently to help the mobile robot to avoid environmental constraint violation and obstacle collision. In other words, the purpose of this work is to check robot traversable directions to guarantee the safety during navigation. Until now, there is a great deal of methods to detect and track environmental obstacles, both static and dynamic objects, using both grid and geometrical approaches [27-38]. To be compatible with the work in the previous section, the grid-based method is presented in this section. For indoor robots, the occupancy grid is a representation of its working environment in which 2D or 3D space is divided into cells, where the resolution of the grid is inversely proportional to the cell size. Each grid cell stores a probabilistic estimate of its state, which may include occupancy, observability, reachability, connectedness, danger, reflectance, and so on. Without loss of generality, a single property is basically presented. The occupancy information should be the most important for the mobile robot planning, which is therefore chosen to be present here. Other properties can be then straightforwardly added to the state of the cell. As presented in [27, 29], the occupancy property of each cell in the occupancy grid is defined by probabilities of two states, including occupied (in short, OCC) and empty (in short, EMP). It is clear that $P\left(C_i^{\mathrm{EMP}}\right) + P\left(C_i^{\mathrm{OCC}}\right) = 1$, in which $P\left(C_i^{\mathrm{EMP}}\right)$ and $P\left(C_i^{\mathrm{OCC}}\right)$ are probabilities of empty and occupied of cell C_i in the grid map, respectively. And, $C_i^{\mathrm{EMP}}, C_i^{\mathrm{OCC}}$ represent empty state and occupied state of cell C_i, respectively. These values depend on the sensing process. By obtaining values of all cells in the grid, an occupancy grid map is then constructed. To do this, a sensor range measurement r is firstly interpreted by a sensor model $p(r|z)$ which is a probability density function. Here, z is the actual distance to the detected object. Then, the conditional probabilities of a cell C_i given a sensor reading r are $P\left(C_i^{\mathrm{OCC}} \mid r\right)$ and $P\left(C_i^{\mathrm{EMP}} \mid r\right)$. After that, the states of cells of the occupancy grid map are updated by using the sensor reading. Generally, the conditional probabilities of all possible world configurations are required to obtain the optimal estimation of the occupancy grid $P\left(C_i \mid r\right)$. In this section, the 2D grid map is considered for the autonomous mobile robot, which includes $m \times n$ cells. It is noted that there are no causal relationships between the occupancy states of different cells. Therefore, the states of each cell are estimated as independent random variables to avoid the combinatorial explosion of grid configurations. The probability of a single cell is also considered by using Bayes theorem, as

$$P\left(C_i^{\mathrm{OCC}} \mid r\right) = \frac{p\left(r \mid C_i^{\mathrm{OCC}}\right) P\left(C_i^{\mathrm{OCC}}\right)}{\sum_{C_i} p\left(r \mid C_i\right) P\left(C_i\right)} \tag{4.10}$$

Here $p\left(r \mid C_i\right)$ involves the range reading with the detection of a single object surface. The sensor model is then

$$p(r \mid z) = p\left(r \mid C_i^{\mathrm{OCC}} \wedge C_k^{\mathrm{EMP}}, \quad k < i\right) \tag{4.11}$$

As presented, the conditional probabilities of all possible world configurations are required to obtain the optimal estimation of the occupancy grid. Here, Kolmogoroff's theorem is used to derive the distributions of $p(r \mid C_i)$:

$$p\left(r \mid C_i^{\mathrm{OCC}}\right) = \sum_{\left\{G_{C_i^{\mathrm{OCC}}}\right\}} \left(p\left(r \mid C_i^{\mathrm{OCC}}, G_{C_i^{\mathrm{OCC}}}\right) \times P\left(G_{C_i^{\mathrm{OCC}}} \mid C_i^{\mathrm{OCC}}\right) \right) \qquad (4.12)$$

Here, $\left\{G_{C_i^{\mathrm{OCC}}}\right\}$ includes all possible configurations $G_{C_i^{\mathrm{OCC}}} = \left(C_1^{\mathrm{OCC}}, \ldots, C_i^{\mathrm{OCC}}, \ldots, C_n^{\mathrm{OCC}}\right)$. In the same way, the probability density of the empty state is

$$p\left(r \mid C_i^{\mathrm{EMP}}\right) = \sum_{\left\{G_{C_i^{\mathrm{EMP}}}\right\}} \left(p\left(r \mid C_i^{\mathrm{EMP}}, G_{C_i^{\mathrm{EMP}}}\right) \times P\left(G_{C_i^{\mathrm{EMP}}} \mid C_i^{\mathrm{EMP}}\right) \right) \qquad (4.13)$$

The configuration probabilities $P\left(G_{C_i^{\mathrm{EMP}}} \mid C_i^{\mathrm{EMP}}\right)$ and $P\left(G_{C_i^{\mathrm{EMP}}} \mid C_i^{\mathrm{EMP}}\right)$ are determined from the individual prior cell state probabilities. These values can be determined by the noninformative or maximum entropy priors, as $P\left(C_i^{\mathrm{OCC}}\right) = P\left(C_i^{\mathrm{EMP}}\right) = \frac{1}{2}$. Besides, the Bayes theorem's sequential updating formulation is used to allow the incremental composition of the sensor information. It is presented as

$$P\left(C_i^{\mathrm{OCC}} \mid \{r\}_{t+1}\right) = \frac{p\left(r_{t+1} \mid C_i^{\mathrm{OCC}}\right) P\left(C_i^{\mathrm{OCC}} \mid \{r\}_t\right)}{\sum_{C_i} p\left(r_{t+1} \mid C_i\right) P\left(C_i \mid \{r\}_t\right)} \qquad (4.14)$$

Here, the current estimate $P\left(C_i^{\mathrm{OCC}} \mid \{r\}_t\right)$ is obtained via observations $\{r_t\} = \{r_1, \ldots, r_t\}$, which is considered as the prior for the next estimation. After $P\left(C_i^{\mathrm{EMP}}\right)$ and $P\left(C_i^{\mathrm{OCC}}\right)$ are obtained, it is necessary to decide whether cell C_i is occupied or empty. It is straightforward to make the decision like the cell C_i is *EMPTY* if $P\left(C_i^{\mathrm{EMP}}\right) > P\left(C_i^{\mathrm{OCC}}\right)$ or the cell C_i is *OCCUPIED* if $P\left(C_i^{\mathrm{EMP}}\right) < P\left(C_i^{\mathrm{OCC}}\right)$; otherwise, the cell C_i is *UNKNOWN*.

For multiple sensors, an occupancy grid integration is normally required only if the sensing system includes various types of sensors. In the case that sensors are same, the simplest way is to merge their data before measuring the state of cells in the occupancy grid map. For example, the robot is equipped with two LiDAR sensors. Their data should be merged as one data set, which is then used for the grid map construction.

Based on the presented occupancy grid, some alternative approaches (called, costmaps) are presented in which the grid values are costs but not probabilities. Here, more complex costs are also added to the costmap. These methods are now popularly used such as [30–35]. Moreover, to improve the performance (related to the response time, range of contexts, tuning, and execution), the layered costmaps for the context-sensitive navigation are presented in [36]. This approach semantically separates the costmap into different layers to process. In this section, environmental modeling is considered by defining cost and risk map as in [36] and is briefly configured as in "*common_params.yaml*," plugged in [25, 26]. The content of "*common_params.yaml*" can be found in Appendix 4.6. Simultaneously, the mobile robot must consider its behavior in both local and global maps. Parameters for local costmap and global costmap are considered in Appendix 4.7. Furthermore, to consider 3D obstacles in working environments, a spatiotemporal voxel layer in [37, 38] is also considered, using the camera point cloud data. An example of this work can be seen in Figure 4.3.

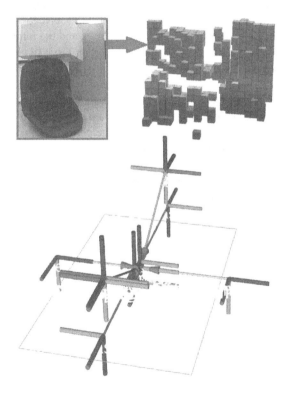

FIGURE 4.3 An example of 3D obstacle detection by using the point cloud data of cameras.

(Photograph by Nguyen Van Toan.)

PATH PLANNING

In the path planning, two problems should be presented, including global planner (to generate the path) and path tracking (to track the generated path). For the global planner, two popular and basic methods are briefly summarized: Dijkstra's algorithm [39, 40] and $A*$ algorithm [41, 42]. Dijkstra's algorithm is presented to find the shortest path from the start to the goal. In this approach, a $n \times n$ matrix $D = [d_{ij}]$ is used to describe the weighted digraph G of n vertices. Here d_{ij} is the length (distance or weight) of the directed edge from the vertex i to the vertex j. If there is no edge from the vertex i to the vertex j, then the value of d_{ij} is very large (or *infinity*). In the case that the edge is available, $d_{ij} > 0$ if $i \neq j$ or $d_{ij} = 0$ if $i = j$. At each stage, the algorithm assigns some vertices as permanent nodes and some other vertices as temporary nodes. In the beginning, the starting vertex (at the current position of the robot) is assigned as the permanent node (values 0), and the remaining $n - 1$ vertices are assigned as temporary nodes (valued *infinity*). In next steps, the value of the vertex j (assigned as a temporary node) is updated by the minimum of (previous value *(j)*, and previous value *(i)* + d_{ij}). Here, i is the latest vertex which is assigned as a permanent node in the previous iteration and d_{ij} is defined as before. This update is applied to all temporary nodes. Then, a vertex with the smallest value is assigned as the permanent node. This procedure is conducted until the target vertex (the target position of the robot) is assigned as the permanent node. The time complexity of Dijkstra's algorithm is $O(n^2)$. To direct toward the goal quicker, a faster search algorithm (called A^*, with the time complexity $O(n)$) is briefly presented, which can be considered as an extension of Dijkstra's algorithm by adding a heuristic value. This value is used to estimate the remaining distance between the examined node and the goal. So, the next permanent node is chosen through the lowest value of the total distance from the starting node to the targeted node $f(node_j) = g(node_j) + h(node_j)$. Here, $g(node_j)$ is the cost from the starting point to the $node_j$ and $h(node_j)$ is the heuristic function to measure the cost from the $node_j$ to the goal. The Euclidean distance can be used to obtain these costs. Similar to Dijkstra's algorithm, the $node_j$ with the lowest $f(node_j)$ is chosen at each step, and its neighbors' costs are also updated for the next iteration.

For the generated path tracking, the pure pursuit algorithm [43, 44] is presented in this section. This approach geometrically determines the curvature that will drive the robot from its current position (x_r, y_r) to a chosen goal point (x_g, y_g) (this goal point is on the generated path). The distance from the (x_r, y_r) to (x_g, y_g) is called *lookahead distance* (in short, *ld*). In this method, an arc is constructed from the (x_r, y_r) to (x_g, y_g) in which the *lookahead distance ld* is considered as the chord length of this arc. It is noted that (x_r, y_r) and (x_g, y_g) are in the global coordinates. Now, (x_{rg}, y_{rg}) is represented for the orthogonal projection of (x_g, y_g) to the local coordinates of the robot. Then, the curvature is determined as

$$
\begin{cases}
\alpha = \arctan 2\left(\dfrac{y_g - y_r}{x_g - x_r}\right) - \varphi \\[2mm]
x_{rg} = ld\sin(\alpha) \\[2mm]
y_{rg} = ld\cos(\alpha) \\[2mm]
R = \dfrac{ld^2}{2y_{rg}}
\end{cases}
\tag{4.15}
$$

Here φ is the angle between the local coordinates of the robot and the global coordinates. The values of x_{rg} and y_{rg} depend on the definition of the local coordinates of the mobile robot. As in the previous chapter (Robot System Analysis), the x axis of the mobile robot is the heading direction (the moving direction of the robot). Therefore, the values of x_{rg} and y_{rg} are obtained in Equation (4.15). Besides, R is the radius of the constructed arc. It is reminded that the goal point (x_g, y_g) is on the generated path that is one *lookahead distance ld* from the current position of the robot. So, to choose a goal point (x_g, y_g), the *lookahead distance ld* should be chosen first. It is straightforward to see that the longer *lookahead distance ld* tends to make less oscillations since it converges to the path more gradually. However, the robot can follow a curvier path if the *lookahead distance ld* is shorter. This observation raises the need to use an optimal (adaptive) *lookahead distance ld*. Until now, a brief summary of the pure pursuit algorithm is presented. To understand more deeply and clearly, readers are encouraged to have a look at [43, 44].

To consider the presented path planning works of the proposed robot system, "*nav2_navfn_planner*" (global planner), "*nav2_regulated_pure_pursuit_controller*" (generated path following), and "*nav2_waypoint_follower*" (waypoint following) are considered. They are also plugged in [25, 26]. Firstly, based on the map and the environmental modeling, a global path is generated from the current position to the goal. This work is conducted by using *NavfnPlanner* in "*nav2_navfn_planner*" [25, 26], as configured in Appendix 4.8. Besides, to follow the generated path, Regulated Pure Pursuit Controller is used via the work in "*nav2_regulated_pure_pursuit_controller*" [25, 26], as configured in Appendix 4.9. Otherwise, the mobile robot is usually requested to travel through a set of waypoints, especially in factories. This observation raises the need to use a waypoint follower. To do this, "*nav2_waypoint_follower*" is considered and configured in Appendix 4.10.

This chapter briefly presents the navigation work of the proposed mobile manipulator system. For more details, an observation on references is necessary. Moreover, to make the robot system work in real-world environments, some other robotic modules should be conducted also, such as packages to activate and obtain data from LiDAR and camera, motor drivers, Wi-Fi module, map switching to help the robot to adapt new working floors, auto docking, and charge dock. For the future work, the performance of the proposed robot system can be improved by using the works in [45, 46].

ABBREVIATIONS

SLAM:	Simultaneous Localization and Mapping
2D:	Two Dimensional
LiDAR:	Light Detection and Ranging
ORB-SLAM3:	An Accurate Open-Source Library for Visual, Visual-Inertial, and Multi-Map SLAM
OpenSLAM:	A Versatile Visual SLAM Framework
RTABMap:	Real-Time Appearance-Based Mapping
Gmapping:	Grid Mapping
MCL:	Monte Carlo Localization
AMCL:	Adaptive Monte Carlo Localization
SIR:	Sampling/Important Re-sampling
Particle MCMC:	Particle Markov chain Monte Carlo
KLD:	Kullback-Leiber Distance
TIM561:	A 2D LiDAR range finder, developed by SICK
NUC:	Next Unit of Computing, designed by Intel
OCC:	Occupied
EMP:	Empty
O(n):	n Operations
O(n^2):	$n \times n$ Operations

REFERENCES

[1] Alexey Merzlyakov, Steve Macenski. *A comparison of modern general-purpose visual SLAM approaches*. In 2021 IEEE/RSJ International Conference on Intelligent Robots and Systems (IROS), Prague, Czech Republic, 27 September–01 October 2021, pp. 9190–9197.

[2] Giorgio Grisetti, Cyrill Stachniss, Wolfram Burgard. *Improved techniques for grid mapping with Rao-Blackwellized particle filters*. IEEE Transactions on Robotics, vol. 23, no. 1, pp. 34–36, 2007.

[3] A ROS Wrapper for OpenSlam's Gmapping. Available at: http://wiki.ros.org/gmapping

[4] Stefan Kohlbrecher, Oskar von Stryk, Johannes Meyer, Uwe Klingauf. *A flexible and scalable SLAM system with full 3D motion estimation*. In Proceedings of 2011 IEEE International Symposium on Safety, Security, and Rescue Robotics, Kyoto, Japan, 01–05 November 2011, pp. 155–160.

[5] Hector Slam Metapackage. Available at: http://wiki.ros.org/hector_slam

[6] Wolfgang Hess, Damon Kohler, Holger Rapp, Daniel Andor. *Real-time loop closure in 2D LiDAR SLAM*. In Proceedings of 2016 IEEE International Conference on Robotics and Automation (ICRA), Stockholm, Sweden, 16–21 May 2016, pp. 1271–1278.

[7] Cartographer Open House. Available at: https://github.com/ros2/cartographer

[8] Steve Macenski, Ivona Jambrecic. *SLAM Toolbox: SLAM for the dynamic world*. Journal of Open Source Software, vol. 6, no. 61, p. 2783, 2021.

[9] SLAM Toolbox Open Source. Available at: https://github.com/SteveMacenski/slam_toolbox

[10] Frank Dellaert, Dieter Fox, Wolfram Burgard, Sebastian Thrun. *Monte Carlo localization for mobile robots*. In Proceedings of 1999 IEEE International Conference on Robotics and Automation (ICRA), Detroit, MI, USA, 10–15 May 1999, pp. 1322–1328.

[11] Donald B. Rubin. *Using the SIR algorithm to simulate posterior distributions*. In Bernardo, J.M., Degroot, M.H., Lindley, D.V. and Smith, A.M. (Eds.), Bayesian Statistics 3: Proceedings of the Third Valencia International Meeting, Clarendon Press, Oxford, 1988, pp. 385–402.

[12] Michael K. Pitt, Neil Shephard. *Filtering and simulation: auxiliary particle filters*. Journal of the American Statistical Association, vol. 95, no. 446, pp. 590–599, 1999.

[13] Walter R. Gilks, Carlo Berzuini. *Following a moving target—Monte Carlo inference for dynamic Bayesian models*. Journal of the Royal Statistical Society Series B (Statistical Methodology), vol. 63, no. 1, pp. 127–146, 2001.

[14] Geir Evensen. *Sequential data assimilation with a nonlinear quasi-geostrophic model using Monte Carlo methods to forecast error statistics*. Journal of Geophysical Research, vol. 99, no. C5, pp. 10143–10162, 1994.

[15] Marco Frei, Hans R. Kunsch. *Bridging the ensemble Kalman and particle filter*. Biometrika, vol. 100, no. 4, pp. 781–800, 2013.

[16] Jing Lei, Peter Bickel. *A moment matching ensemble filter for nonlinear non-Gaussian data assimilation*. Monthly Weather Review, vol. 139, no. 12, pp. 3964–3973, 2011.

[17] Mark Briers, Arnaud Doucet, Simon Maskell. *Smoothing algorithms for state-space models*. Annals of the Institute of Statistical Mathematics, vol. 62, Article number 61, pp. 61–89, 2010.

[18] Paul Fearnhead, David Wyncoll, Jonathan Tawn. *A sequential smoothing algorithm with linear computational cost*. Biometrika, vol. 97, no. 2, pp. 447–464, 2010.

[19] Christophe Andrieu, Arnaud Doucet, Roman Holenstein. *Particle Markov chain Monte Carlo methods.* Journal of the Royal Statistical Society Series B (Statistical Methodology), vol. 72, no. 3, pp. 269–342, 2010.

[20] Dieter Fox. *KLD-sampling: adaptive particle filters.* In Advances in Neural Information Processing Systems 14 [Neural Information Processing Systems: Natural and Synthetic, NIPS 2001], Vancouver, British Columbia, Canada, 3–8 December 2001, pp. 713–720.

[21] Norman L. Johnson, Samuel Kotz, Narayanaswamy Balakrishnan. Continuous univariate distributions, vol. 1, John Wiley & Sons, New York, 1994.

[22] Dieter Fox, Wolfram Burgard, Frank Dellaert, Sebastian Thrun. *Monte Carlo localization: efficient position estimation for mobile robots.* In Proceedings of the Sixteenth National Conference on Artificial Intelligence and the Eleventh Innovation Applications of Artificial Intelligence Conference Innovative Applications or Artificial Intelligence, Orlando, Florida, USA, 18–22 July 1999, pp. 343–349.

[23] James Carpenter, Peter Clifford, Paul Fearnhead. *An improved particle filter for non-linear problems.* IEE Proceedings—Radar, Sonar and Navigation, vol. 146, no. 1, pp. 2–7, 1999.

[24] A ROS2 Message Converter. Available at: https://github.com/mehmetkillioglu/ros2_message_converter/tree/master

[25] Steve Macenski, Francisco Martín, Ruffin White, Jonatan Ginés Clavero. *The Marathon 2: A navigation system.* In 2020 IEEE/RSJ International Conference on Intelligent Robots and Systems (IROS), Las Vegas, NV, USA, 25–29 October 2020, pp. 2718–2725.

[26] Navigation 2 System. Available at: https://github.com/ros-planning/navigation2

[27] Larry Matthies, Alberto Elfes. *Integration of sonar and stereo range data using a grid-based representation.* In Proceedings of 1988 IEEE International Conference on Robotics and Automation (ICRA), Philadelphia, PA, USA, 24–29 April 1988, pp. 727–733.

[28] Nguyen Van Toan, Minh Hoang Do, Jaewon Jo. *MoDeT: A low-cost obstacle tracker for self-driving mobile robot navigation using 2D-laser scan.* Industrial Robot: The International Journal of Robotics Research and Application, vol. 49, no. 6, pp. 1032–1041, 2022. https://doi.org/10.1108/IR-12-2021-0289.

[29] Alberto Elfes. *Occupancy Grid: a stochastic spatial representation for active robot perception.* In Proceedings of the Sixth Conference on Uncertainty in Artificial Intelligence (UAI 1990), Cambridge, MA, 27–29 July 1990, pp. 136–146.

[30] Emrah Akin Sisbot, Luis F. Marin-Urias, Rachid Alami, Thierry Simeon. *A human aware mobile robot motion planner.* IEEE Transactions on Robotics, vol. 23, no. 5, pp. 874–883, 2007.

[31] Dave Ferguson, Maxim Likhachev. *Efficiently using cost maps for planning complex maneuvers.* In Proceedings of the Workshop on Planning with Cost Maps, IEEE International Conference on Robotics and Automation, Pasadena, CA, USA, 19–23 May 2008.

[32] Brian P. Gerkey, Motilal Agrawal. *Break on through: tunnel-based exploration to learn about outdoor terrain.* In Proceedings of the Workshop on Planning with Cost Maps, IEEE International Conference on Robotics and Automation, Pasadena, CA, USA, 19–23 May 2008.

[33] Rachel Kirby, Reid Simmons, Jodi Forlizzi. *COMPANION: a constraint-optimizing method for person-acceptable navigation.* In Proceedings of the 18th IEEE Symposium on Robot and Human Interactive Communication (Ro-Man), Toyama, Japan, 27 September–02 October 2009, pp. 607–612.

[34] Mikael Svenstrup, Soren Tranberg, Hans Jorgen Andersen, Thomas Bak. *Pose estimation and adaptive robot behavior for human-robot interaction.* In Proceedings of the IEEE International Conference on Robotics and Automation (ICRA), Kobe, Japan, 12–17 May 2009, pp. 3571–3576.

[35] Leonardo Scandolo, Thierry Fraichard. *An anthropomorphic navigation scheme for dynamic scenarios.* In Proceedings of the IEEE International Conference on Robotics and Automation (ICRA), Shanghai, China, 09–13 May 2011, pp. 809–814.

[36] David V. Lu, Dave Hershberger, William D. Smart. *Layered costmaps for context-sensitive navigation.* In 2014 IEEE/RSJ International Conference on Intelligent Robots and Systems (IROS 2014), Chicago, IL, USA, 14–18 September 2014, pp. 709–715.

[37] Steve Macenski, David Tsai, Max Feinberg. *Spatio-temporal voxel layer: a view on robot perception for the dynamic world.* International Journal of Advanced Robotics Systems, March 2020, doi:10.1177/1729881420910530.

[38] Spatio-temporal Voxel Layer Package. Available at: https://github.com/SteveMacenski/spatio_temporal_voxel_layer

[39] Donald E. Knuth. *A generation of Dijkstra's algorithm.* Information Processing Letters, vol. 6, no. 1, pp. 1–5, 1977.

[40] Thomas H. Cormen, Charles E. Leiserson, Ronald L. Rivest, Clifford Stein. Introduction to algorithms (Second ed.), Section 24.3: Dijkstra's algorithm. MIT Press and McGraw-Hill, Cambridge, pp. 595–601, 2001.

[41] Rina Dechter, Judea Pearl. *Generalized best-fit search strategies and the optimality of A*.* Journal of ACM, vol. 32, no. 3, pp. 505–536, 1985.

[42] A. R. Soltani, H. Tawfik, J. Y. Goulermas, T. Fernando. *Path planning in construction sites: performance evaluation of the Dijkstra, A*, and GA search algorithms.* Advanced Engineering Informatics, vol. 16, no. 4, pp. 291–303, 2002.

[43] R. Craig Coulter. *Implementation of the Pure Pursuit path tracking algorithm.* Technical Report CMU-RI-TR-92-01, The Robotics Institute, Carnegie Mellon University Pittsburg, Pennsylvania, January 1992.

[44] Kresimir Petrinec, Zdenko Kovacic, Alessandro Marozin. *Simulator of multi-AGV robotic industrial environments.* In IEEE International Conference on Industrial Technology, Maribor, Slovenia, 10–12 December 2003, pp. 979–983.

[45] Dechao Chen, Shuai Li, Qing Wu. *A novel supertwisting zeroing neural network with application to mobile robot manipulators.* IEEE Transactions on Neural Networks and Learning Systems, vol. 32, no. 4, pp. 1776–1787, 2021.

[46] Bor-Sen, Yueh-Yu Tsai, Min-Yen Lee. *Robust decentralized formation tracking control for stochastic large-scale biped robot team system under external disturbance and communication requirements.* IEEE Transactions on Control of Network Systems, vol. 8, no. 2, pp. 654–666, 2021.

APPENDIX

4.1: The connection between the user interface and the core of SLAM mode (cartographer), Navigation mode (navigation2), and robot driver.

```
import rclpy, os, signal, subprocess
terminate = False
def signal_handling (signum, frame):
    global terminate
    terminate = True
signal.signal(signal.SIGINT, signal_handling)
if __name__ == '__main__':
    rclpy.init()
    node = rclpy.create_node('robot_actions')
    rate = node.create_rate(1)
    # Path to execution files
    home_dir = os.getenv("HOME")
    RESTART_DIR = '%s/catkin_ws/src/robot_lib/launch/robot_start.
            launch.xml'%home_dir
    NAV_DIR = '%s/catkin_ws/src/robot_lib/launch/nav.launch.xml'%home_dir
    SLAM_DIR = '%s/catkin_ws/src/robot_lib/launch/slam_2d.launch.
            xml'%home_dir
    # Restarting the robot
    ro_pid = subprocess.Popen('ros2 launch RESTART_DIR', shell=True,
            preexec_fn=os.setsid)
    global mode; mode = "SLAM"; prev_mode = "SLAM"
    while rclpy.ok():
        if terminate == True:
            node.get_logger().info("Terminate the process …")
            break
        if mode == "NAV":
            nav_pid = subprocess.Popen('ros2 launch NAV_DIR',
                shell=True, preexec_fn=os.setsid)
            node.get_logger().info("[MAPPING] Robot is under NAV
                mode …")
            current_mode = mode
            while current_mode == mode:
                if terminate == True: break
            os.killpg(os.getpgid(nav_pid.pid), signal.SIGTERM);
            rate.sleep()
            prev_mode = "NAV"
        elif mode == "SLAM":
            slam_pid = subprocess.Popen('ros2 launch SLAM_DIR',
                        shell=True, preexec_fn=os.setsid)
            node.get_logger().info("[MAPPING] Robot is under SLAM
                mode …")
            current_mode = mode
            while current_mode == mode:
                if terminate == True: break
            os.killpg(os.getpgid(slam_pid.pid), signal.SIGTERM);
            rate.sleep()
            prev_mode = "SLAM"
```

```
    elif mode == "RESTART":
        node.get_logger().info("[MAPPING] Rebooting the robot …")
        os.killpg(os.getpgid(ro_pid.pid), signal.SIGTERM);
        rate.sleep()
        ro_pid = subprocess.Popen('ros2 launch RESTART_DIR',
                shell=True, preexec_fn=os.setsid)
        mode = prev_mode
    else:
        node.get_logger().info("[MAPPING] System error …")
        break
os.killpg(os.getpgid(ro_pid.pid), signal.SIGTERM); rate.sleep()
```

4.2: The content of *"slam_2d.launch.xml"'* in **Appendix 4.1**.

```xml
<launch>
<arg name="resolution" default="0.02"/>
<arg name="configuration_directory" default="$(find cartographer_node)/
                                 launch/"/>
<arg name="configuration_basename" default="slam_2d.lua"/>
<node pkg="cartographer_ros" exec="cartographer_node"
            name="cartographer_node" output="screen">
    <param name="configuration_directory" value="$(var
            configuration_directory)"/>
    <param name="configuration_basename" value="$(var
            configuration_basename)"/>
</node>
<node pkg="cartographer_ros" exec="cartographer_occupancy_grid_node"
            name="cartographer_occupancy_grid_node" output="screen"/>
    <param name="resolution" value="$(var resolution)"/>
</node>
</launch>
```

4.3: The content of *"slam_2d.lua"* in **Appendix 4.2**.

```lua
include "map_builder.lua"
include "trajectory_builder.lua"
RID = os.getenv("ROS_HOSTNAME")
options = {
map_builder = MAP_BUILDER, trajectory_builder = TRAJECTORY_BUILDER,
map_frame = "map",
tracking_frame = string.format( "%s/base_link",RID),
published_frame = string.format( "%s/odom",RID),
odom_frame = string.format( "%s/odom",RID),
provide_odom_frame = false, publish_frame_projected_to_2d = false,
use_pose_extrapolator = on,
use_odometry = true,use_nav_sat = false, use_landmarks = false, num_
laser_scans = 1,
num_multi_echo_laser_scans = 0, num_subdivisions_per_laser_scan = 10,
num_point_clouds = 0,
lookup_transform_timeout_sec = 0.2, submap_publish_period_sec = 0.3,
pose_publish_period_sec = 5e-3, trajectory_publish_period_sec = 30e-3,
rangefinder_sampling_ratio = 1., odometry_sampling_ratio = 1.,fixed_
                         frame_pose_sampling_ratio = 1.,
imu_sampling_ratio = 1., landmarks_sampling_ratio = 1.,
}
```

```
MAP_BUILDER.use_trajectory_builder_2d = true
TRAJECTORY_BUILDER_2D.num_accumulated_range_data = 10
TRAJECTORY_BUILDER_2D.min_range = 0.1
TRAJECTORY_BUILDER_2D.max_range = 25.
TRAJECTORY_BUILDER_2D.missing_data_ray_length = 5.
TRAJECTORY_BUILDER_2D.use_imu_data = false
TRAJECTORY_BUILDER_2D.use_online_correlative_scan_matching = true
TRAJECTORY_BUILDER_2D.motion_filter.max_angle_radians = math.rad(0.1)
POSE_GRAPH.constraint_builder.min_score = 0.65
POSE_GRAPH.constraint_builder.global_localization_min_score = 0.7
return options
```

4.4: The function to save the map.

```
from ros2_message_converter import message_converter
map_data = OccupancyGrid()
node.create_subscription(OccupancyGrid, "map", callback)
def callback(msg):
      try:
              global map_data
              map_data = msg
      except:
              node.get_logger().info("[MAP_DATA] Cannot receive map data …")
btn_map_save.clicked.connect(save_map)
def save_map():
      try:
              # Path Creation
              map_path = os.getenv("MAP_PATH")
              map_dir = "%s/map"%map_path
              if not os.path.exists(map_dir): os.mkdir(map_dir)

              # Save Map topics
              map_json = message_converter.convert_ros_message_to_
              dictionary(map_data)
              map_file = open("%s/map.json"%map_dir, 'wt')
              map_file.write(map_json); map_file.close()

              # Save Map Images
              RID = os.getenv("ROS_HOSTNAME")
              subprocess.Popen('ros2 run map_server map_saver --ros-args -p
              occ:=50 -p free:=40 -p f:= %s/map'%(map_dir), shell=True)
      except:
              node.get_logger().info("[SAVE_MAP] Errors happened during
              saving the map …")
```

4.5: A set of parameters used for AMCL in this chapter.

```
amcl:

  ros__parameters:                    min_particles: 100
    use_sim_time: false               odom_frame_id: "odom"
    alpha1: 0.01                      pf_err: 0.1
    alpha2: 0.04                      pf_z: 0.5
    alpha3: 0.04                      recovery_alpha_fast: 0.0
    alpha4: 0.002                     recovery_alpha_slow: 0.0
```

```
base_frame_id: "base_footprint"        resample_interval: 1
beam_skip_distance: 0.5                robot_model_type: "differential"
beam_skip_error_threshold: 0.9         save_pose_rate: 0.5
beam_skip_threshold: 0.3               sigma_hit: 0.2
do_beamskip: false                     tf_broadcast: true
global_frame_id: "map"                 transform_tolerance: 2.0
lambda_short: 0.1                      update_min_a: 0.05
laser_likelihood_max_dist: 2.0         update_min_d: 0.06
laser_max_range: 25.0                  z_hit: 0.95
laser_min_range: 0.1                   z_max: 0.05
laser_model_type: "likelihood_field"   z_rand: 0.5
max_beams: 180                         z_short: 0.05
max_particles: 2500
                                       scan_topic: scan
```

4.6: The common parameters to define the cost and risk map.

```
global_frame: map                 obstacle_min_range: 0.1     # meters
robot_base_frame: base_link       max_obstacle_height: 2.0    # meters
update_frequency: 5.0             raytrace_max_range: 10.0     # meters
publish_frequency: 0.0           raytrace_min_range: 0.0
static_map: true                 footprint: [[0.54226,   0.400], [0.54226,
rolling_window: false                        -0.400], [-0.507, -0.400],
                                             [-0.507,  0.400]]
#START VOXEL STUFF               footprint_padding: 0.0
map_type: voxel                  inflation_radius: 0.7
origin_z: 0.0                    cost_scaling_factor: 20.0
z_resolution: 0.2               lethal_cost_threshold: 60
z_voxels: 10                     observation_sources: base_scan
unknown_threshold: 15 # voxel height  base_scan: {data_type: LaserScan,
mark_threshold: 0     # voxel height            expected_update_rate: 0.4,
#END VOXEL STUFF                               observation_persistence: 0.0,
transform_tolerance: 0.2  # seconds            marking: true, clearing: true,
obstacle_max_range: 6.0 # meters               max_obstacle_height: 0.4,
                                               min_obstacle_height: 0.08}
```

4.7: Parameters for local and global costmap.

```
local_costmap:                    global_costmap:
  local_costmap:                    global_costmap:
    ros__parameters:                  ros__parameters:
      update_frequency: 30.0            update_frequency: 5.0
      publish_frequency: 2.0            publish_frequency: 2.0
      global_frame: odom                global_frame: map
      robot_base_frame: base_link       robot_base_frame: base_link
      rolling_window: true              use_sim_time: True
      width: 6                          robot_radius: 0.68
      height: 6                         resolution: 0.02
      resolution: 0.1                   track_unknown_space: true
      plugins: ["voxel_layer",          plugins: ["static_layer",
      "inflation_layer"]                "obstacle_layer",
      inflation_layer:                          "inflation_layer"]
```

```
      plugin: "nav2_costmap_2d::
        InflationLayer"
      cost_scaling_factor: 3.0
      inflation_radius: 0.55
    voxel_layer:
      plugin: "nav2_costmap_2d::
        VoxelLayer"
      enabled: True
      publish_voxel_map: True
      origin_z: 0.0
      z_resolution: 0.05
      z_voxels: 16
      max_obstacle_height: 2.0
      mark_threshold: 0
      observation_sources: scan
      scan:
        topic: /scan
        max_obstacle_height: 2.0
        clearing: True
        marking: True
        data_type: "LaserScan"
        raytrace_max_range: 3.0
        raytrace_min_range: 0.0
        obstacle_max_range: 2.5
        obstacle_min_range: 0.0
    static_layer:
      map_subscribe_transient_
        local: True
    always_send_full_costmap: True
```

```
    obstacle_layer:
      plugin: "nav2_costmap_2d::
        ObstacleLayer"
      enabled: True
      observation_sources: scan
      scan:
        topic: /scan
        max_obstacle_height: 2.0
        clearing: True
        marking: True
        data_type: "LaserScan"
        raytrace_max_range: 3.0
        raytrace_min_range: 0.0
        obstacle_max_range: 2.5
        obstacle_min_range: 0.0
    static_layer:
      plugin: "nav2_costmap_2d::
        StaticLayer"
      map_subscribe_transient_
        local: True
    inflation_layer:
      plugin: "nav2_costmap_2d::
        InflationLayer"
      cost_scaling_factor: 3.0
      inflation_radius: 0.55

    always_send_full_costmap: True
```

4.8: The configuration of the global path generator in this chapter.

```
planner_server:
  ros__parameters:
    expected_planner_frequency: 20.0
    use_sim_time: True
    planner_plugins: ["GridBased"]
    GridBased:
      plugin: "nav2_navfn_planner/NavfnPlanner"
      tolerance: 0.5
      use_astar: false
      allow_unknown: true
```

4.9: The configuration of the path follower in this chapter.

```
controller_server:

  ros__parameters:
    use_sim_time: True
    controller_frequency: 15.0
    min_x_velocity_threshold: 0.001
    min_y_velocity_threshold: 0.5
    min_theta_velocity_threshold:
      0.001
    progress_checker_plugin:
      "progress_checker"
```

```
    max_linear_accel: 1.2
    max_linear_decel: 1.2
    lookahead_dist: 0.9
    min_lookahead_dist: 0.6
    max_lookahead_dist: 1.8
    lookahead_time: 2.0
    rotate_to_heading_angular_
      vel: 0.8
    transform_tolerance: 0.1
    use_velocity_scaled_
      lookahead_dist: false
```

```
goal_checker_plugin:
  "goal_checker"
controller_plugins: ["FollowPath"]

progress_checker:
  plugin: "nav2_controller::Simp
    leProgressChecker"
  required_movement_radius: 0.5
  movement_time_allowance: 20.0
goal_checker:
  plugin: "nav2_controller::Simp
    leGoalChecker"
  xy_goal_tolerance: 0.15
  yaw_goal_tolerance: 0.05
  stateful: True
FollowPath:
  plugin: "nav2_regulated_pure_
    pursuit_controller::
  RegulatedPurePursuitController"
  desired_linear_vel: 0.8
```

```
min_approach_linear_velocity:
  0.03
use_approach_linear_velocity_
  scaling: true
max_allowed_time_to_
  collision: 1.0
use_regulated_linear_
  velocity_scaling: true
use_cost_regulated_linear_
  velocity_scaling:
                      false
regulated_linear_scaling_min_
  radius: 0.9
regulated_linear_scaling_min_
  speed: 0.25
use_rotate_to_heading: true
rotate_to_heading_min_angle:
  0.785
max_angular_accel: 1.7
cost_scaling_dist: 0.3
cost_scaling_gain: 1.0
inflation_cost_scaling_
  factor: 3.0
```

4.10: The configuration of the waypoint follower in this chapter.

```
waypoint_follower:
  ros__parameters:
    loop_rate: 20
    stop_on_failure: false
    waypoint_task_executor_plugin: "wait_at_waypoint"
    wait_at_waypoint:
      plugin: "nav2_waypoint_follower::WaitAtWaypoint"
      enabled: True
      waypoint_pause_duration: 200
```

Manipulator Manipulation

5

Nguyen Van Toan

To manipulate the targeted objects during the execution of the designed tasks, the manipulator on the mobile robot plays a crucial role. This chapter briefly presents the manipulator manipulation work of the proposed robot system, regarding the formulation of the manipulator modeling, trajectory planning, and control in the software system.

MANIPULATOR MODELING

System modeling is the first work which should be considered before motion planning and control. In the chapter Robot System Analysis, transformation coordinates as well as kinematic and dynamic relationships among joints and links of the manipulator are theoretically presented. However, to describe a robot in a software system, it is necessary to use a standardized file format [1]. In robot operating system (ROS), the data structure of a robot model is represented by the unified robot description format (URDF) [2, 3] and the semantic robot description format (SRDF) [4]. The URDF file is used to describe properties relevant to robots, such as kinematic constraints, three-dimensional geometric representation, joint limits, courser-grained collision geometry, geometric visualization meshes, sensors, and dynamic properties of links and joints (such as mass, moments of inertia, and velocity limits). These parameters are integrated to synchronize the motions of the real robot and its visualization. Here, robots are restricted to include rigid links only. By contrast, the SRDF file does not describe physical properties of robots, while it provides reconfigurable semantic meta-data of the robot model. The information in SRDF file complements the URDF and is useful for motion

DOI: 10.1201/9781003352426-5

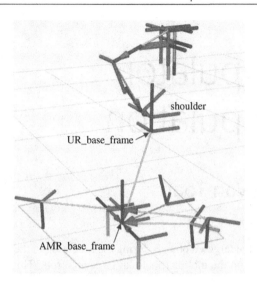

FIGURE 5.1   Transformation coordinates of the mobile manipulator.
(Photograph by Nguyen Van Toan.)

planning, such as specifying joint groups (actuated or passive), default robot configurations, and which set of links constructs poses of the end effector.

As analyzed in Chapter 3, Robot System Analysis, an example of transformation coordinates of the proposed robot system is shown in Figure 5.1, in which coordinates of both the mobile robot and the manipulator are described. However, the mobile robot and the manipulator are designed to work independently. Namely, the mobile robot is stationary when the manipulator is working, and vice versa. Therefore, the manipulator manipulation is presented separately. It is noticed that the kinematic and dynamic constraints of the mobile robot and the manipulator must be considered if their kinematics are synchronized (or if they work together to manipulate the targeted objects).

Kinematic parameters of the proposed manipulator (with respect to the UR_base_frame) are presented in Table 5.1. It is noted that the manipulator is a Universal UR5 model. Currently, Moveit2 [5] is popularly used for manipulators. Besides, Universal ROS driver [6] is also constructed for universal robot arms. They are therefore leveraged for the manipulator-manipulation work of the proposed robot system. In analogy of the definition in [5, 6], joints' names of the manipulator are called: *shoulder, upper_arm, forearm, wrist_1, wrist_2,* and *wrist_3*. Here, the tool frame is not included.

TABLE 5.1   Kinematic parameters of the manipulator (UR5, home configuration)

	X	Y	Z	ROLL	PITCH	YAW
shoulder	0.0	0.0	0.089159	0.0	0.0	0.0
upper_arm	0.0	0.0	0.0	90°	0.0	0.0
Forearm	−0.425	0.0	0.0	0.0	0.0	0.0
wrist_1	−0.39225	0.0	0.10915	0.0	0.0	0.0
wrist_2	0.0	−0.09465	0.0	90°	0.0	0.0
wrist_3	0.0	0.0823	0.0	90°	0.0	0.0

TABLE 5.2   Dynamic parameters of the manipulator (UR5)

		CENTER OF MASS			SHAPE	
	MASS	X	Y	Z	RADIUS	LENGTH
shoulder	3.7	0.0	0.00193	−0.02561	0.06	0.178
upper_arm	8.393	0.0	−0.024201	0.2125	0.054	0.425
forearm	2.275	0.0	0.0265	0.11993	0.04	0.39225
wrist_1	1.219	0.0	0.110949	0.01634	0.045	0.095
wrist_2	1.219	0.0	0.0018	0.11099	0.045	0.085
wrist_3	0.1879	0.0	0.001159	0.0	0.0375	0.305

In addition, the dynamic parameters of the manipulator (UR5 robot) are presented in Table 5.2.

As presented, these kinematic and dynamic parameters are used to construct the manipulator in the software system, under the URDF and SRDF file formats, as configured in Appendix 5.1.

Figure 5.2 is a description of the manipulator (UR5) in rviz, by loading the urdf and srdf files of the manipulator, including kinematic and dynamic relationships, and visualization of links and joints.

TRAJECTORY PLANNING AND CONTROL

There is a great deal of trajectory planning methods for the manipulator, and they have their own advantages and disadvantages. Toward a broad approach, the open motion planning library (OMPL) is considered for the proposed robot system. The OMPL contains implementations of many state-of-the-art planning algorithms, such as Probabilistic Roadmap

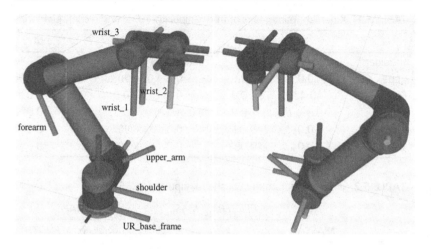

FIGURE 5.2   A visualization of manipulator (UR5).

(Photograph by Nguyen Van Toan.)

Method (PRM) [7], SPArse Roadmap Spanner algorithm (SPARS) [8], Rapidly exploring Random Trees (RRT) [9], Expansive Space Trees (EST) [10], Kinodynamic Planning by Interior-Exterior Cell Exploration (KPIECE) [11], Path-Directed Subdivision Trees (PDST) [12], Search Tree with Resolution Independent Density Estimation (STRIDE) [13], Fast Marching Tree algorithm (FMT*) [14], and many more. A complete list of available planners in current OMPL can be found in [15, 16]. Together with OMPL, the KDL (kinematics and dynamics library) is also used to convert from a Cartesian space to joint configuration space via some numerical techniques. This library is for forward and inverse kinematics for generic arms as well as custom plugins, and its details can be found in [17]. It is noted that trajectories are specified as a set of waypoints, which consist of positions, velocities, and accelerations. After joint-space trajectories are obtained, a controller should be used to reach them at a specific time instants. Here, the joint trajectory controller in [18] is chosen, which is templated to work with multiple trajectory representations as well as multiple hardware interface types. The interpolation strategies in this controller can be linear (only consider the position), cubic (consider position and velocity), or quintic (consider position, velocity, and acceleration), in which desired positions are simply forwarded to the joints if only the position is specified. Otherwise, the position and velocity trajectory following error is mapped to the velocity (or effort) commands through a PID (proportional integral derivative) loop.

As mentioned, Moveit2 [5] and Universal Robot driver [6] are leveraged for motion planning and control in this section. First of all, *ur.srdf. xacro* is defined as the semantic description of the manipulator and *ur.urdf. xacro* is defined as the description of the manipulator, based on kinematic and dynamic parameters in the previous section. In *ur.urdf.xacro*, following parameters should be mentioned:

```
robot_ip = "192.168.3.10"
joint_limits_parameters_file = "~/ur_description/joint_
limits.yaml"
kinematics_parameters_file = "~/ur_description/default_
kinematics.yaml"
physical_parameters_file = "~/ur_description/physical_
parameters.yaml"
```

The *robot_ip* is the Internet Protocol (IP) address of the manipulator to communicate with other computers and controllers in the network. In the proposed robot system, the IP address of the manipulator is configured as "*192.168.3.10*." The *joint_limits.yaml* defines the limit of every joints of the manipulator. The *default_kinematics.yaml* and *physical_parameters. yaml* present kinematic and dynamic relationships, respectively, following the forms in the previous section. Now, the planning configuration should be implemented. In this section, an *ompl* planner is used. Its configuration is presented in Appendix 5.2. Here, *ompl_planning_yaml* is the path to *ompl_planning.yaml*, which contains the custom configuration of the planner. Next, the trajectory execution configuration should be concerned, which can be found in Appendix 5.3. Here, *controller.yaml* contains the controller, which is used to follow the path. And the *FollowJointTrajectory* is chosen. Finally, the actual move group server should include the robot description, the robot description semantic, the kinematic description, the planning configuration, the trajectory execution, the controller, and planning scene monitor. The form should be: *robot_description* contains the content of *ur.urdf.xacro*, *robot_description_semantic* contains the content of *ur.srdf.xacro*, and *robot_description_kinematics* contains the content of *kinematics.yaml* which defines the kinematic solver for the manipulator; here *KDLKinematics* is used:

```
# Start the actual move_group node/action server
move_group_node = Node(
    package="moveit_ros_move_group",
    executable="move_group",
    output="screen",
    parameters=[
```

```
    robot_description,
    robot_description_semantic,
    robot_description_kinematics,
    ompl_planning_pipeline_config,
    trajectory_execution,
    moveit_controllers,
    planning_scene_monitor_parameters,
  ],
)
```

This chapter briefly presented the manipulator modeling, planning, and control by leveraging Moveit2 library [5] and Universal Robot driver [6]. For more details, observation on [5, 6] is necessary.

ABBREVIATIONS

ROS: Robot Operating System
URDF: Unified Robot Description Format
SRDF: Semantic Robot Description Format
MoveIt: MoveIt Motion Planning Framework
OMPL: Open Motion Planning Library
ORM: Probabilistic Roadmap Method
SPARS: SPArse Roadmap Spanner
RRT: Rapidly Exploring Random Trees
EST: Expansive Space Trees
KPIECE: Kinodynamic Planning by Interior-Exterior Cell Exploration
PDST: Path-Directed Subdivision Trees
STRIDE: Search Tree with Resolution Independent Density Estimation
FMT*:   Fast Marching Tree
KDL: Kinematics and Dynamics Library
PID: Proportional Integral Derivative
IP: Internet Protocol

REFERENCES

[1] David Coleman, Ioan Sucan, Sachin Chitta, Nikolaus Correll. *Reducing the barrier to entry of complex robotic software: a MoveIt! Case study.* Journal of Software Engineering for Robotics, vol. 1, no. 1, pp. 1–14, 2014.

[2] A Tutorial on Unified Robot Description Format (URDF). Available at: https://docs.ros.org/en/foxy/Tutorials/Intermediate/URDF/URDF-Main.html

[3] A C++ Parser for the Unified Robot Description Format (URDF). Available at: http://wiki.ros.org/urdf

[4] Semantic Robot Description Format. Available at: http://wiki.ros.org/srdf

[5] MoveIt Motion Planning Framework for ROS2. Available at: https://github.com/ros-planning/moveit2

[6] Universal Robots ROS2 Driver. Available at: https://github.com/UniversalRobots/Universal_Robots_ROS2_Driver

[7] Lydia E. Kavraki, Petr Svestka, Jean-Claude Latombe, Mark H. Overmars. *Probabilistic roadmaps for path planning in high-dimensional configuration spaces.* IEEE Transactions on Robotics and Automation, vol. 12, no. 4, pp. 566–580, 1996.

[8] Andrew Dobson, Athanasios Krontiris, Kostas E. Bekris. *Parse roadmap spanners.* In Proceedings of the 14th Workshop on the Algorithmic Foundations of Robotics (WAFR), Cambridge, Massachusetts, USA, 13–15 June 2012, pp. 279–296.

[9] Steven M. LaValle, James J. Kuffner. *Randomized kinodynamic planning.* The International Journal of Robotics Research, vol. 20, no. 5, pp. 378–400, 2001.

[10] David Hsu, Jean-Claude Latombe, Rajeev Motwani. *Path planning in expansive configuration spaces.* International Journal of Computational Geometry and Applications, vol. 9, nso. 4–5, pp. 495–512, 1999.

[11] Ioan A. Sucan, Lydia E. Kavraki. *A sampling-based tree planner for systems with complex dynamics.* IEEE Transactions on Robotics, vol. 28, no. 1, pp. 116–131, 2012.

[12] Andrew M. Ladd, Lydia E. Kavraki. *Motion planning in the presence of drift, underactuation and discrete system changes.* Robotics: Science and Systems I, Cambridge, Massachusetts, USA, 8–11 June 2005, pp. 233–241.

[13] Bryant Gipson, Mark Moll, Lydia E. Kavraki. *Resolution independent density estimation for motion planning in high-dimensional spaces.* In 2013 IEEE International Conference on Robotics and Automation (ICRA), Karlsruhe, Germany, 06–10 May 2013, pp. 2429–2435.

[14] Lucas Janson, Edward Schmerling, Ashley Clark, Marco Pavone. *Fast marching tree: a fast marching sampling-based method for optimal motion planning in many dimensions.* The International Journal of Robotics Research, vol. 34, no. 7, pp. 883–921, 2015.

[15] Ioan A. Sucan, Mark Moll, Lydia E. Kavraki. *The open motion planning library.* IEEE Robotics & Automation Magazine, vol. 19, no. 4, pp. 72–82, 2012.

[16] OMPL Available Planners. Available at: http://ompl.kavrakilab.org/planners.html

[17] KDL: Kinematics and Dynamics Library. Available: http://www.orocos.org/kdl.html

[18] Controller for Executing Joint-Space Trajectories. Available at: http://wiki.ros.org/joint_trajectory_controller

APPENDIX

5.1 Kinematic and dynamic parameters of the manipulator.

```
inertia_parameters:                  kinematics:
    shoulder_mass: 3.7000              shoulder:
        .                                x: 0.0
        .                                y: 0.0
    wrist_3_mass: 0.1879                 z: 0.089159
                                         roll: 0.0
links:                                   pitch: 0.0
    shoulder:                            yaw: 0.0
        radius: 0.06                   upper_arm:
        length: 0.178                    x: 0.0
            .                            y: 0.0
            .                            z: 0.0
    wrist_3:                             roll: 1.570796327
        radius: 0.0375                   pitch: 0.0
        length: 0.0305                   yaw: 0.0
                                             .
center_of_mass:                              .
    shoulder_cog:                            .
        x: 0.0                               .
        y: 0.00193
        z: -0.02561                    wrist_3:
            .                            x: 0.0
            .                            y: 0.0823
    wrist_3_cog:                         z: 0.0
        x: 0.0                           roll: 1.57
        y: 0.001159                      pitch: 0.0
        z: 0.0                           yaw: 0.0
```

5.2: The configuration of the motion planner for the manipulator.

```
# Planning Configuration
ompl_planning_pipeline_config = {
    "move_group": {
        "planning_plugin": "ompl_interface/OMPLPlanner",
        "request_adapters": """default_planner_request_adapters/
                            AddTimeOptimalParameterization
                                default_planner_request_adapters/
                                FixWorkspaceBounds
                                default_planner_request_adapters/
                                FixStartStateBounds
                                default_planner_request_adapters/
                                FixStartStateCollision
                            default_planner_request_adapters/
                            FixStartStatePathConstraints""",
        "start_state_max_bounds_error": 0.1,
    }
}

ompl_planning_pipeline_config["move_group"].update(ompl_planning_yaml)
```

5.3: The configuration of the trajectory execution for the manipulator.

```
# Trajectory Execution Configuration
moveit_controllers = {
    "moveit_simple_controller_manager": controllers_yaml,
    "moveit_controller_manager":
        "moveit_simple_controller_manager/MoveItSimpleControllerManager",
}

trajectory_execution = {
    "moveit_manage_controllers": False,
    "trajectory_execution.allowed_execution_duration_scaling": 1.2,
    "trajectory_execution.allowed_goal_duration_margin": 0.5,
    "trajectory_execution.allowed_start_tolerance": 0.01,
}

planning_scene_monitor_parameters = {
    "publish_planning_scene": True,
    "publish_geometry_updates": True,
    "publish_state_updates": True,
    "publish_transforms_updates": True,
    "planning_scene_monitor_options": {
        "name": "planning_scene_monitor",
        "robot_description": "robot_description",
        "joint_state_topic": "/joint_states",
        "attached_collision_object_topic": "/move_group/
          planning_scene_monitor",
        "publish_planning_scene_topic": "/move_group/
          publish_planning_scene",
        "monitored_planning_scene_topic": "/move_group/
          monitored_planning_scene",
        "wait_for_initial_state_timeout": 10.0,
    },
}
```

Robot Perception

6

Nguyen Van Toan

The robot perception is a vital system which endows the robot with the ability to perceive and understand about the targeted objects and the working environment. In this chapter, some perception works are presented to help the proposed mobile manipulator to change its working floors successfully. As analyzed in the chapter Task-Oriented Robot System Proposal, those perception works should be: the elevator button detection, the elevator button light-up checking, the elevator door status checking, the human detection and the floor number recognition.

ELEVATOR BUTTON DETECTION

To operate the elevator by manipulating its buttons, the elevator button recognition should be a fundamental step and can be considered as a visual object detector. Obviously, it is overwhelmingly challenging to use classical image processing algorithms to handle this task since there are many types of elevator panels and buttons installed in different environments. With great practical success, deep learning architectures such as Region-based Convolutional Neural Network (RCNN) [1–3], You Only Look Once (YOLO) [4], and single-shot multibox detector (SSD) [5] should be now most effective for this issue. Be inspired by these methods, some elevator button recognizers are presented in [6–12] and shown reasonable results. Then, to enhance the performance regarding to the robustness and the accuracy, a combination of the faster RCNN architecture and an optical character recognition (OCR) network [13] is proposed in [14]. In analogy with the typical faster RCNN architecture, the framework for the elevator button recognition in [14] is basically

DOI: 10.1201/9781003352426-6

composed by a region proposal network (RPN), an object classifier, and a bounding refinement branch. These branches share the same convolutional feature map, which is generated by a certain Feature Extractor [15]. First, the RPN takes an image of any size as the input to generate a set of rectangular object proposals (here, each region proposal is attached with an objectness score), whose purpose is to tell the object detector where (regions of interest, ROIs) in the raw image should be looked. Based on objectness scores, a non-max suppression is implemented to choose high-score proposals. They are then passed to the Box Prediction (including Classification and Location Refinement), where those ROIs' locations are used to find out their corresponding feature patches in the shared feature map. For further classification and location refinement, those patches are cropped and resized to a batch of fixed-length Abstract Feature Proposal. Besides, an OCR branch is added for the labeling task, which shares the Abstract Feature Proposal with the Classification and Location Refinement branches. This OCR branch is used to distinguish elevator buttons based on characters included in their areas. In the final step, the network gives labels and locations of elevator buttons in the input image. The algorithm was tested with 630 elevator button panels, 6869 buttons. The average precision of the detection is 0.946. After that, its alternative improvements are presented in [16]. Thus, there exist various algorithms for the elevator button recognition. Each algorithm has its own advantages and disadvantages since they have tried to solve different problems in the elevator button recognition. To robustly recognize various types of elevator buttons in various modern multi-floor buildings, the works in [14] and [16] have been proven to provide the better performance in comparison with other methods. They are therefore chosen for the elevator button recognition work of the proposed robot system. For more details in theory, an observation on Refs. [14] and [16] is recommended.

The main function to implement the elevator button recognition based on the works in [14] and [16] is briefly presented in Appendix 6.1. Here, *ButtonDetector* and *CharacterRecognizer* are button region detection and character recognition classes. First of all, *ButtonDetector* is used to detect elevator button regions in the input image. These regions are then cropped to become inputs of *CharacterRecognizer* to recognize characters inside them. It is noticed that the output *button_status* is the result of the button light-up status-checking function, which is named *light_check*. The content of the function *light_check* is presented in the next section. The main process of *ButtonDetetor* can be seen in Appendix 6.2. Here, the *detection_graph.pb* is the pretrained model of the elevator button detector, located in the folder *~/model*, and can be downloaded by using the link in [17]. The program is built based on Tensorflow 1.14.0, using CPU only, Intel NUC Core i7, without GPU. As mentioned, after elevator buttons are detected, they are cropped to become inputs of *Character Recognizer* to recognize characters inside their regions. To avoid content duplication, the detail of *CharacterRecognizer* class should be found in Floor Number Recognition Section because that module is also used for another task in this chapter. By sharing a module for several tasks with similar purposes, the software system is optimized both in the computation and the structural complexity.

Figure 6.1 is a result obtained by the elevator button recognition. The upper part of Figure 6.1 includes the button recognition results (score,

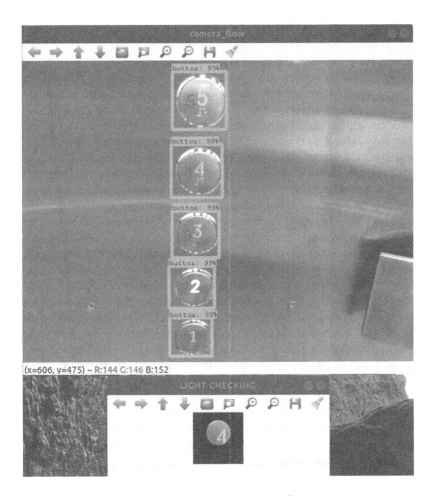

FIGURE 6.1 Elevator button recognition, the fourth floor is the target.

(Photograph by Nguyen Van Toan.)

boundary, and the number inside the button). The lower part is the result of the button-light status checking. Here, only targeted button (in Figure 6.1, the button number four is the target) is considered for the button-light status checking. After the targeted elevator button is recognized, the pixel coordinates and the depth information of its center are obtained to become the input of the manipulator planner.

BUTTON LIGHT-UP STATUS CHECKING

After elevator buttons are pressed successfully, characters inside them should be lighted on. This information is the useful signal not only for persons, but also for the robot to discover whether or not the targeted button is manipulated successfully. In this section, a solution is implemented to check the elevator button light-up status. It is clear that methods should be chosen so that they are as simple as possible while targeted problems are still solved effectively. Here, to check the elevator button light-up status, the change of the color of characters inside the targeted button is estimated before and after it is manipulated. The checking procedure should be as follows: (step 6A1) detect and crop the targeted elevator button, (step 6A2) convert the RGB color of the cropped image into the HSV coordinates, (step 6A3) compare the HSV color space with a threshold range ([lower, upper]) and extract qualified regions (their HSV color belongs to the threshold range), (step 6A4) find contours of those regions, and measure their total areas, and (step 6A5) compare the total area of those contours with a threshold to decide this button is light-up or not. The threshold in (step 6A5) can be absolute or relative. Namely, the total area of contours is compared with a specific threshold, or the proportion of the total area of contours to the total area of the cropped button is compared with a ratio threshold (its value belongs to [0–1]). The relative threshold is used because the distance from the robot to the elevator button panel is different case by case, depending on the offset of the actual and nominal poses of the mobile robot. Meanwhile, the area of the button in the image is bigger if the mobile robot is closer to the elevator button panel.

Through the presented procedure, the elevator button recognition (as presented in the previous section), the HSV color transformation and the contour extraction should be concerned to check the elevator button light-up status. Since only the targeted elevator button is manipulated by the manipulator, only its light-up status is checked. Therefore, the ROI is uniquely the area inside contours of the targeted button's characters. To begin with, the targeted elevator button is firstly detected and cropped from the input image. It then plays a role as the input of the HSV (hue-saturation-value) transformation. Shortly, the HSV is known as an alternative representation of the RGB color space, in which the color information is represented by hue and saturation values. Hue is the color portion of the HSV system to represent basic colors. It shall be signified as a point in a 360-degree color circle (ranges from 0 to 360 degrees). Its value is determined by the dominant wavelength in the spectral distribution of light wavelengths. Saturation (the vibrancy of the color) is directly connected to the intensity of the color (range of gray in the color space). It is normally represented in terms of percentage, from 0 to 100 percent. Finally, value refers to the brightness of the color, which is represented as percentage (ranges from 0 to 100 percent, where 0 percent represents black and 100 percent presents the brightest). This approach is similar to the human perception of colors and commonly used in computer graphics applications. It is noted that the HSV color system is not device independent, but it is only defined relative to the RGB color (a nonlinear transformation of the RGB color space). As presented in [18, 19], the HSV values can be transformed from the RGB space easily, as

$$H = \arccos\left(\frac{R - \frac{1}{2}G - \frac{1}{2}B}{\sqrt{(R-G)^2 + (R-B)(G-B)}}\right) \text{ if } B \le G$$

$$H = 360 - \arccos\left(\frac{R - \frac{1}{2}G - \frac{1}{2}B}{\sqrt{(R-G)^2 + (R-B)(G-B)}}\right) \text{ if } G < B \qquad (6.1)$$

$$S = \frac{\max(R, G, B) - \min(R, G, B)}{\max(R, G, B)}$$

$$V = \frac{\max(R, G, B)}{255}$$

As mentioned above, after the targeted elevator button is detected and cropped from the input RGB image, it is transformed to the HSV coordinates. Then, a filter is conducted through the range of pixel values in the HSV color map. The HSV threshold range must be chosen via the color of elevator button numbers when they are turned on. In the next steps, the filtered image is used to extracted contours of ROIs (as a curve joining all the continuous points with same color or intensity), and their areas are measured to decide whether the targeted button status is "*on*" or "*off*" by comparing the total area of ROIs with an area threshold. Again, the absolute area threshold (the total area of

ROIs) or the relative area threshold (the ratio of the total area of ROIs to the total area of the targeted button) can be used. It is reminded that the contour extraction is one of the important steps to obtain the area of ROIs, which is conducted by using the algorithm in [20]. This method considers only digital binary pictures sampled at points of rectangular grids and derives a sequence of the coordinates or the chain codes from the border between a connected component of 1 pixel (one component) and a connected component of 0 pixel (background or hole). With this algorithm, the surroundness relation among two types of borders (outer borders and hole borders) can be extracted, which corresponds to the surroundness relation among connected components.

The elevator button recognition work in the presented procedure of the button light-up status checking can be found in the previous section. In this section, the HSV transformation and the contour extraction are two main functions that should be concerned. The *light_check* function is conducted by using OpenCV, as in Appendix 6.3. Here, white color is the color of characters inside elevator buttons after they are pressed successfully. The status *"on"* represents for the button light up (already pressed successfully) and *"off"* represents for the button light down (still not pressed successfully).

Figures 6.2 and 6.3 are obtained from the elevator button light-up status checking. The left part of Figure 6.2 and the upper part of Figure 6.3 present the light-up status of targeted buttons before they are pressed. By

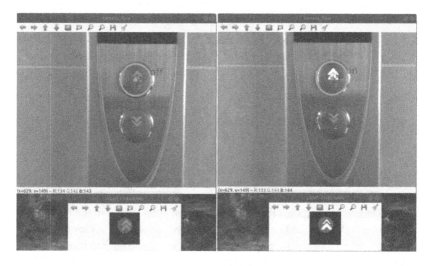

FIGURE 6.2 Outside of elevator: button recognition and light-up status checking.

(Photograph by Nguyen Van Toan.)

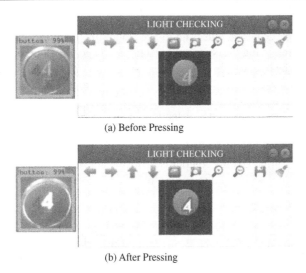

(a) Before Pressing

(b) After Pressing

FIGURE 6.3   Inside of elevator: button recognition and light-up status checking. (Photograph by Nguyen Van Toan.)

contrast, other parts present the light-up status of targeted buttons after they are pressed successfully. Figure 6.2 is outside of the elevator and Figure 6.3 is inside of the elevator. In Figure 6.2, the *"Up"* button is the target. In Figure 6.3, the button number four is the target.

ELEVATOR DOOR STATUS CHECKING

Another important perception work is to check the elevator door status (*"opened"* or *"closed"*) to help the robot to decide when it can get in (or get out of) the elevator. Based on our survey, there is always a warning marker on the elevator door. By recognizing this marker, the robot is aware of whether the elevator door is *"closed"* or *"opened."* A simple idea for this work can be described as: if the marker is disappeared during a sample time, then the elevator door status should be *"opened."* Otherwise, the door status should be *"closed."* To recognize the warning marker on the elevator door, its specific features should be considered, such as its shape (rectangle) and colors (yellow, white, red, and blue). The procedure should be started by a transformation from the input RGB image to the HSV color. After that, some image processing operations are applied, such as Gaussian blur, erosion, and dilation. Here, Gaussian blur is an image-smoothing method which is highly effective in removing Gaussian noise from the image. Erosion operation is to erode away the boundaries of foreground object which is useful for removing small white noises. And, dilation operation is to continuously increase the boundaries of regions of foreground pixels, which is useful in joining broken parts of an object. Next, edge detection and the contour extraction are conducted sequentially. Then, the perimeter of obtained contours is calculated and passed to a polyline simplification procedure. Until now, the shape of the object is determined. If the contour is a satisfied rectangle, then the condition regarding colors is checked to ensure that the detected rectangle is the warning marker.

Some main processes should be concerned such as HSV transformation, the contour extraction, the edge detection, and the contour approximation. In this section, only the edge detection and the contour approximation are briefly introduced since the HSV transformation and the contour extraction are same as in the previous section. The edge detection is conducted by using the method in [21, 22], which is also known to many as the Canny Edge detector or the optimal detector. The algorithm is proposed to satisfy three main criteria: low error rate, good localization (the offset of detected edge pixels and real edge pixels should be minimized), and minimal response (only one detector response per edge). Basically, the process of this method includes four steps: noise filter (using Gaussian filter), image intensity gradient exploration (find the gradient strength and direction, rounded to one of four possible angles 0, 45, 90 or 135 degrees), non-maximum suppression (to consider only thin lines as candidate edges), and hysteresis (reject or accept pixel as an edge by using upper and lower thresholds). Apart from the edge detection, another process should be concerned in this section, which is the contour approximation, whose purpose is to simplify a polyline by reducing its vertices. This issue is overcome by the Ramer-Douglas-Peucker algorithm [23, 24], in which a similar curve with fewer points for a given polyline is found. Here, the simplified curve consists of a subset of points that define the original polyline. First, the start and end points of a given curve are used to draw the shortest line (the reference line). Then, the farthest point (this point is on the given curve) from the reference line is determined. If the distance from the farthest point to the reference line is less than the threshold, all vertices between the start and end points are neglected, and make the curve a straight line. By contrast, the farthest point is considered as a reference point (to make new reference lines and iterate the checking process) if its distance from the reference line is bigger than the threshold. For a polyline consisting of n vertices (and $n - 1$ segments), the time complexity should be $\mathbf{O}(n^2)$ in the worst case and $\mathbf{O}(n \cdot \mathbf{log}(n))$ in the best case.

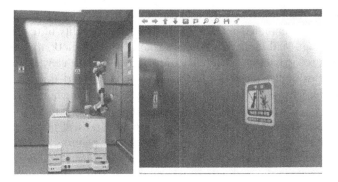

FIGURE 6.4   Elevator door status checking (*"opened"* or *"closed"*).
(Photograph by Nguyen Van Toan.)

In summary, to check the status (*"opened"* or *"closed"*) of the elevator door, the HSV transformation [18, 19], the contour extraction [20], the edge detection [21, 22], and the contour approximation [23, 24] should be mainly concerned. And, the procedure for the elevator warning marker detection can be simply described by two steps: (step 6B1) find qualified rectangles in the input image by using introduced algorithms and (step 6B2) check colors inside those qualified rectangles to identify which is the correct warning marker. Color checking is used as in Button Light-up Status Checking Section. It is noted that used thresholds in algorithms can be absolute or relative (depending on their depth information since the distance from the robot to the elevator door is different case by case). The warning marker detection is implemented by using OpenCV, as presented in Appendix 6.4. Figure 6.4 is an example of the warning marker detection. In this case, the marker appears in the camera frame and is recognized and bounded by a green box. Therefore, the elevator door status is currently *"closed."*

HUMAN DETECTION

Human safety is the first priority when the robot system is working. To avoid threatenings to persons, the robot must be aware of their appearance in its current workspace, especially in the human-crowded areas. It is known that the proposed robot system is requested to work in such kind of environments because there are many persons in modern multi-floor buildings. And, they usually use elevators to move to different floors. Therefore, the human detection should be a crucial ability of the proposed robot system. Within the procedure to change its work-floor, the robot system must recognize the appearance of persons inside the elevator to behave safe. For example, the robot just gets in the elevator if there is no person inside. Similar to the elevator button recognition, the human detection is also categorized into the visual object detection task in the computer vision field. Here, all instances of human beings present in an image are searched and compared with known templates or patterns of people. It is also overwhelmingly challenging to use classical image processing algorithms to handle this task efficiently and effectively. Therefore, the use of deep learning architectures with great practical success for this problem should be a reasonable approach. So far, there is a great deal of human detection algorithms, as in [25–28]. To achieve good performance with a modest hardware system, the MobileNets [25] is chosen for the proposed robot system. Basically, MobileNets use depth-wise separable convolutions to construct light-weight deep neural networks, which focus on optimizing for the latency and the network size. In MobileNets, the depth-wise convolution applies a single filter to each input channel. In comparison with standard convolution, the depth-wise convolution is much more efficient. Unfortunately, it does not combine input channels to create new feature but only filters them. Therefore, an additional layer (called point-wise convolution) is added to compute a linear combination of the output of the depth-wise convolution to generate these new features. In other words, depth-wise separable convolutions are built up by depth-wise convolutions and point-wise convolutions. In fact, these two layers are separated. As a result, the computation and model size are reduced considerably. Besides, all layers of MobileNet are followed by a batch-norm [29] and rectified linear unit (ReLU nonlinearity) with the exception of the final fully connected layer. In discoveries, the accuracy of MobileNet is a bit smaller than standard convolutions while its computation is less than between 8 or 9 times (since 3×3 depth-wise separable convolutions are used in MobileNet architectures).

In this section, the *DNN* module in OpenCV is used to implement the MobileNetSSD algorithm, using CPU only, Intel NUC Core i7, without GPU. The main function to implement the human detection is briefly shown in Appendix 6.5, where *MobileNetdeploy.prototxt* and *MobileNetSSD. caffemodel* define pretrained network layers and the caffe model of the MobileNetSSD architecture, respectively. They can be downloaded by using links in [30, 31]. Normally, the MobileNetSSD is used for the object detection. Here, the person is considered as an object, and its defined ID is 15 in the *classNames*. Figure 6.5 is an example of the human detection inside the elevator.

FIGURE 6.5   Checking the appearance of persons inside the elevator. (Photograph by Nguyen Van Toan.)

FLOOR NUMBER RECOGNITION

Commonly, the robot has to share the elevator with some others (persons or other robots). As a result, the elevator usually stops and opens at random floors (targeted floors of some others). Normally, LCD panels are installed inside and outside of the elevator to display its current floor number and current moving direction. After entering inside elevator and pressing the targeted button successfully, the manipulator must adjust its configuration to locate the camera in front of the LCD panel to obtain the displayed information, whose purpose is to help the robot to be aware of whether or not it arrived its targeted floor. In this section, an OCR network is used to clear the mentioned issue. As presented in Elevator Button Detection Section, this OCR module is same as in elevator button recognition. Therefore, the *CharacterRecognizer* class in Section 6.1 can be referred here. By sharing a module for several

FIGURE 6.6   LCD number recognition to check the targeted floor.
(Photograph by Nguyen Van Toan.)

tasks with similar purposes, the software system is optimized both in computation and structural complexity. Similarly, the program is built based on Tensorflow 1.14.0, using CPU only, Intel NUC Core i7, without GPU. The main function of the *CharacterRecognizer* can be found in Appendix 6.6, where the *ocr_graph.pb* is pretrained model of the optimal character recognition, which can be downloaded by using the link in Ref. [32].

Figure 6.6 is an example of LCD number recognition. In this figure, the number four is recognized which means that the elevator is currently arriving (or already arrived) the fourth floor.

ABBREVIATIONS

RCNN: Region-based Convolutional Neural Network
YOLO: You Only Look Once
SSD: Single-Shot Multibox Detector
OCR: Optical Character Recognition
RPN: Region Proposal Network
ROI: Region of Interest
NMS: Non-Max Suppression
CPU: Central Processing Unit
NUC: Next Unit of Computing, designed by Intel

GPU:	Graphics Processing Unit
RGB:	Red-Green-Blue
HSV:	Hue-Saturation-Value
OpenCV:	Open Source Computer Vision Library
O(n^2):	$n \times n$ Operations
O($n.\log(n)$):	$n \times \log(n)$ Operations
Log(n):	Logarithm of n
ReLU:	Rectified Linear Unit
DNN:	Deep Neural Networks
ID:	Identification Number
LCD:	Liquid Crystal-Display

REFERENCES

[1] Shaoqing Ren, Kaiming He, Ross Girshick, Jian Sun. *Faster R-CNN: towards real-time object detection with region proposal networks.* In Proceedings of the 28th International Conference on Neural Information Processing Systems, Montreal, Canada, 7–12 December 2015, pp. 91–99.

[2] Ross Girshick, Jeff Donahue, Trevor Darrell, Jitendra Malik. *Rich feature hierarchies for accurate object detection and semantic segmentation.* In 2014 IEEE Conference on Computer Vision and Pattern Recognition, Columbus, OH, USA, 23–28 June 2014, pp. 580–587.

[3] Jifeng Dai, Yi Li, Kaiming He, Jian Sun. *R-FCN: object detection via region-based fully convolutional networks.* In Proceedings of the 30th International Conference on Neural Information Processing Systems, Barcelona, Spain, 5–10 December 2016, pp. 379–387.

[4] Joseph Redmon, Santosh Divvala, Ross Girshick, Ali Farhadi. *You only look once: unified, real-time object detection.* In 2016 IEEE Conference on Computer Vision and Pattern Recognition (CVPR), Las Vegas, NY, USA, 27–30 June 2016, pp. 779–788.

[5] Wei Liu, Dragomir Anguelov, Dumitru Erhan, Christian Szegedy, Scott Reed, Cheng-Yang Fu, Alexander C. Berg. *SSD: single shot multibox detector.* In European Conference on Computer Vision, Amsterdam, Netherlands, 11–14 October 2016, pp. 21–37.

[6] Yann Lcun, Leon Bottou, Yoshua Bengio, Patrick Haffner. *Gradient-based learning applied to document recognition.* Proceedings of the IEEE, vol. 86, no. 11, pp. 2278–2324, 1998.

[7] Ellen Klingbeil, Carpenter Blake, Russakovsky Olga, Ng Andrew. *Autonomous operation of novel elevators for robot navigation.* In 2010 IEEE International Conference on Robotics and Automation (ICRA), Anchorage, AK, USA, 03–07 May 2010, pp. 751–758.

[8] Heon-Hui Kim, Dae-Jin Kim, Kwang-Hyun Park. *Robust elevator button recognition in the presence of partial occlusion and cluster by*

specular reflections. IEEE Transactions on Industrial Electronics, vol. 59, no. 3, pp. 1597–1611, 2012.

[9] Wan Nur Farhanah Wan Zakaria, Mohd Razali Daud, Saifudin Razali, Mohammad Fadhil. *Elevators external button recognition and detection for vision-based system.* In Proceeding of International Conference on Electrical Engineering, Computer Science and Informatics (EECSI 2014), Yogyakarta, Indonesia, 20–21 August 2014, pp. 265–269.

[10] Ali A. Abdulla, Hui Liu, Norbert Stoll, Kerstin Thurow. *A new robust method for mobile robot multifloor navigation in distributed life science laboratories.* Journal of Control Science and Engineering, vol. 2016, Articles ID 3589395, 2016, pp. 1–17.

[11] Zijian Dong, Delong Zhu, Max Q.-H. Meng. *An autonomous elevator button recognition system based on convolutional neural networks.* In 2017 IEEE International Conference on Robotics and Biomimetics (ROBIO), Macau, Macao, 06–08 December 2017, pp. 2533–2539.

[12] Delong Zhu, Tingguang Li, Danny Ho, Max Q.-H. Meng. *Deep reinforcement learning supervised autonomous exploration in office environments.* In 2018 IEEE International Conference on Robotics and Automation (ICRA), Brisbane, QLD, Australia, 21–25 May 2018, pp. 7548–7555.

[13] Oriol Vinyals, Lukasz Kaiser, Terry Koo, Slav Petrov, Ilya Sutskever, Geoffrey Hinton. *Grammar as a foreign language.* In Proceedings of the 28th International Conference on Neural Information Processing Systems, Montreal, Canada, 7–12 December 2015, pp. 2773–2781.

[14] Delong Zhu, Tingguang, Danny Ho, Tong Zhou, Max Q.-H. Meng. *A novel OCR-RCNN for elevator button recognition.* In 2018 IEEE/RSJ International Conference on Intelligent Robots and Systems (IROS), Madrid, Spain, 01–05 October 2018, pp. 3626–3631.

[15] Kaiming He, Xiangyu Zhang, Shaoqing Ren, Jian Sun. *Deep residual learning for image recognition.* In 2016 IEEE Conference on Computer Vision and Pattern Recognition (CVPR), Las Vegas, NV, USA, 27–30 June 2016, pp. 770–778.

[16] Delong Zhu, Zhe Min, Tong Zhou, Tingguang Li, Max Q.-H. Meng. *An autonomous eye-in-hand robotic system for elevator button operation based on deep recognition network.* IEEE Transactions on Instrumentation and Measurement, vol. 70, Art no. 2504113, pp. 1–13, 2021. doi:10.1109/TIM.2020.3043118.

[17] A Pre-trained Model for the Elevator Button Recognition. Available at: https://drive.google.com/file/d/1ucwhSJN8VTtPj99i0YJ3EBrYHqOKhRnn/view?usp=sharing

[18] Zhong Liu, Weihai Chen, Yuhua Zou, Cun Hu. *Region of interest extraction based on HSV color space.* In IEEE 10th International Conference on Industrial Informatics, Beijing, China, 25–25 July 2012, pp. 481–485.

[19] Martin Loesdau, Sébastien Chabrier, Alban Gabillon. *Hue and saturation in the RGB color space.* In Elmoataz A., Lezoray O., Nouboud F., Mammass D. (Eds.), Image and Signal Processing. ICISP 2014. Lecture Notes in Computer Science, vol. 8509. Springer, Cham, 2014.

[20] Satoshi Suzuki, Keiichi Abe. *Topological structural analysis of digitized binary images by border following.* Computer Vision, Graphics, and Image Processing, vol. 30, no. 1, pp. 32–46, 1985.

[21] John Canny. *A computational approach to edge detection.* IEEE Transactions on Pattern Analysis and Machine Intelligence, vol. PAMI-8, no. 6, pp. 679–698, 1986.

[22] Cai-xia Deng, Gui-bin Wang, Xin-rui Yang. *Image Edge Detection Algorithm Based on Improved Canny Operator.* In Proceedings of the 2013 International Conference on Wavelet Analysis and Pattern Recognition, Tianjin, China, 14–17 July 2013, pp. 168–172.

[23] Urs Ramer. *An iterative procedure for the polygonal approximation of plane curves.* Computer Graphics and Image Processing, vol. 1, no. 3, pp. 244–256, 1972.

[24] David H. Douglas, Thomas K. Peucker. *Algorithms for the reduction of the number of points required to represent a digitized line or its caricature.* The Canadian Cartograher, vol. 10, no. 2, pp. 112–122, 1973.

[25] Andrew G. Howard, Menglong Zhu, Bo Chen, Dmitry Kalenichenko, Weijun Wang, Tobias Weyand, Marco Andreetto, Hartwig Adam. *MobileNets: efficient convolutional neural networks for mobile vision applications.* ArXiv, 2017.

[26] Denis Tome, Chris Russell, Lourdes Agapito. *Lifting from the deep: convolutional 3D pose estimation from a single image.* In Proceedings of the IEEE Conference on Computer Vision and Pattern Recognition, Honolulu, HI, USA, 21–26 July 2017, pp. 5689–5698.

[27] Georgios Pavlakos, Xiaowei Zhou, Konstantinos G. Derpanis, Kostas Daniilidis. *Coarse-to-fine volumetric prediction for single-image 3D human pose.* In Proceedings of 2017 IEEE Conference on Computer Vision and Pattern Recognition (CVPR), Honolulu, HI, USA, 21–26 July 2017, pp. 1263–1272.

[28] Dario Pavllo, Christoph Feichtenhofer, David Grangier, Michael Auli. *3D human pose estimation in video with temporal convolutions and semi-supervised training.* In Proceedings of the Conference on Computer Vision and Pattern Recognition (CVPR), Long Beach, CA, USA, 16–20 June 2019, pp. 7753–7762.

[29] Sergey Ioffe, Christian Szegedy. *Batch Normalization: accelerating deep network training by reducing internal covariate shift.* In Proceedings of the 32nd International Conference on Machine Learning, Lille, France, 6–11 July 2015, pp. 448–456.

[30] A Pre-trained Network Layers of MobileNetSSD Architecture. Available at: https://drive.google.com/file/d/1vQAB3g5cSy0Q7xqq039bac7RkXfqhdmY/view?usp=sharing

[31] A Caffe Model of MobileNetSSD Architecture. Available at: https://drive.google.com/file/d/1uWowi7uj8cotCoNvRYCpa0tgTuuoDy19/view?usp=sharing

[32] A Pre-trained Model of the Optimal Character Recognition. Available at: https://drive.google.com/file/d/1qwW1_omzdWaOwKcLfTaUQ9v6Xc-ATO-t/view?usp=sharing

APPENDIX

6.1: The main function to implement the button recognition.

```python
import cv2
detector = ButtonDetector()              # button region detection class
recognizer = CharacterRecognizer()       # character recognition class

def button_recognition
(image, targeted_floor):                 # "targeted_floor" is the target floor
                                         # of the robot
    h_f,w_f = image.shape[:2]            # the height and the width of the
                                         # image, respectively
    desired_button_pos = []
    button_status = ''
    boxes, scores, _ = detector.predict(image)

    for box, score in zip(boxes, scores):
        if score < 0.5: continue

        y_min = int(box[0] * h_f)
        x_min = int(box[1] * w_f)
        y_max = int(box[2] * h_f)
        x_max = int(box[3] * w_f)

        button_image = image[y_min: y_max, x_min: x_max]
        button_image = cv2.resize(button_image, (180, 180))
        button_text, button_score = recognizer.predict(button_image)

        if(button_text == targeted_floor):
            coor = [(int)((x_min+x_max)/2),(int)((y_min+y_max)/2)]
            desired_button_pos.append(coor)
            button_status = light_check(button_image)  # check the
                                         # button light-up function
    return desired_button_pos, button_status
```

6.2: Main functions of *ButtonDetetor* in **Appendix *6.1*.**

```python
import tensorflow as tf
# load frozen graph
graph_path = '~/models/detection_graph.pb'
detection_graph = tf.Graph()
with detection_graph.as_default():
    od_graph_def = tf.GraphDef()
    with tf.gfile.GFile(graph_path, 'rb') as fid:
        serialized_graph = fid.read()
        od_graph_def.ParseFromString(serialized_graph)
        tf.import_graph_def(od_graph_def, name='')
session = tf.Session(graph=detection_graph)

# prepare input and output request
class_num = 1
output = []
img_input = detection_graph.get_tensor_by_name('image_tensor:0')
output.append(detection_graph.get_tensor_by_name('detection_boxes:0'))
```

```
output.append(detection_graph.get_tensor_by_name('detection_scores:0'))
output.append(detection_graph.get_tensor_by_name('detection_classes:0'))
output.append(detection_graph.get_tensor_by_name('num_detections:0'))

def predict(image_np):
        img_in = numpy.expand_dims(image_np, axis=0)
        boxes, scores, classes, num = session.run(output, feed_dict={img_
        input: img_in})
        boxes, scores, classes, num = [numpy.squeeze(x) for x in [boxes,
        scores, classes, num]]
        return boxes, scores, num
```

6.3: The button light up checking *(light_check)* function.

```
import cv2, imutils, numpy
def light_check(img):
        total_area = 0
        button_status = ''
        # BEGIN HSV
        try:
              hsv = cv2.cvtColor(img, cv2.COLOR_BGR2HSV)
              sensitivity = 30
              lower_white = numpy.array([0,0,255-sensitivity])
              upper_white = numpy.array([255,sensitivity,255])
              max_cnt = []
              yeo_mask = cv2.inRange(hsv, lower_white, upper_white)
              yeo_cnts = cv2.findContours(yeo_mask.copy(), cv2.RETR_EXTERNAL,
                                cv2.CHAIN_APPROX_SIMPLE)
              cnts = imutils.grab_contours(yeo_cnts)
              cnt_area = [0.0]*len(cnts)
              contourLength  = len(cnts)
              if contourLength < 1:
                      button_status = 'off'
                      pass
              else:
                      for i in range(contourLength):
                      cnt_area[i] = cv2.contourArea(cnts[i])
                              total_area += cnt_area[i]
                              #cv2.drawContours(img, cnt_area[i], -1, (0, 0,
                              # 255), 2)
                      max_cnt.append(cnts[cnt_area.index(max(cnt_area))])
                      if total_area > 80:
                      #if max(cnt_area) > 80:
                              button_status = 'on'
                              cv2.drawContours(img, max_cnt, -1, (0, 0, 255), 2)
                      else:
                              button_status = 'off'
        except:
                pass
        return button_status
```

6.4: The warning marker detection for the elevator door status checking.

```
import cv2, cv_bridge, imutils, numpy
def warning_marker_detection(image):
        # Detect the warning marker area
        cnts = None
```

```
img_out = image
gray = cv2.cvtColor(image,cv2.COLOR_BGR2GRAY)
gray = cv2.GaussianBlur(gray, (3, 3), 0)
kernel = numpy.ones((3,3),numpy.uint8)
erosion = cv2.erode(gray,kernel,iterations = 2)
dilation = cv2.dilate(erosion,kernel,iterations = 2)
edges = cv2.Canny(dilation, 50, 200, 255)
cnts = cv2.findContours(edges.copy(), cv2.RETR_EXTERNAL, cv2.
CHAIN_APPROX_SIMPLE)
cnts = imutils.grab_contours(cnts)
cnts = sorted(cnts, key=cv2.contourArea, reverse=True)
displayCnt = None
cX = 0
cY = 0

# loop over the contours
for c in cnts:
        # approximate the contour
        peri = cv2.arcLength(c, True)
        approx = cv2.approxPolyDP(c, 0.02 * peri, True)
        if len(approx) == 4:
            displayCnt = approx
            output = four_point_transform(image, displayCnt.
            reshape(4, 2))
                    # BEGIN HSV
            hsv = cv2.cvtColor(output, cv2.COLOR_BGR2HSV)
            #BEGIN FILTER
            lower_yellow = numpy.array([ 20, 190, 20])  # for yellow
            upper_yellow = numpy.array([30, 255, 255])
            yeo_mask = cv2.inRange(hsv, lower_yellow, upper_yellow)
            (_, yeo_cnts, _) = cv2.findContours(yeo_mask.copy(),
            cv2.RETR_EXTERNAL, cv2.CHAIN_APPROX_SIMPLE)
        # Loop through all of the contours, and get their areas
            contourLength  = len(yeo_cnts)
            if contourLength >= 1:
                M = cv2.moments(c)
                if(M["m00"] != 0):
                    cX = int(M["m10"] / M["m00"])
                    cY = int(M["m01"] / M["m00"])
                    cv2.drawContours(img_out, [c], -1, (0, 255,
                    0), 2)
                else:
                    cX = 0
                    cY = 0
                break
    return cX, cY   # if cX and cY are equal zeros  => cannot detect
    # the marker
```

6.5: The main function to implement the human detection using MobileNetSSD.

```
# caffe mobilenet model
net = cv2.dnn.readNetFromCaffe('model/MobileNetdeploy.prototxt', 'model/
MobileNetSSD.caffemodel')
classNames = {0: 'background', 15: 'person'}

def human_locations(image, thr=0.5):
        frame_resized = cv2.resize(image, (300,300))
```

```
blob = cv2.dnn.blobFromImage(frame_resized, 0.007843, (300, 300),
(127.5, 127.5, 127.5), False)
net.setInput(blob)
#Prediction of network
detections = net.forward()
#Size of frame resize (300x300)
cols = frame_resized.shape[1]
rows = frame_resized.shape[0]
output = []

for i in range(detections.shape[2]):
    confidence = detections[0, 0, i, 2] #Confidence of prediction
    if confidence > thr: # Filter prediction
        class_id = int(detections[0, 0, i, 1]) # Class label

        # Object location
        xLeftBottom = int(detections[0, 0, i, 3] * cols)
        yLeftBottom = int(detections[0, 0, i, 4] * rows)
        xRightTop   = int(detections[0, 0, i, 5] * cols)
        yRightTop   = int(detections[0, 0, i, 6] * rows)

        # Factor for scale to original size of image
        heightFactor = image.shape[0]/300.0
        widthFactor = image.shape[1]/300.0
        # Scale object detection to image
        xLeftBottom = max(0, int(widthFactor * xLeftBottom))
        yLeftBottom = max(0, int(heightFactor * yLeftBottom))
        xRightTop   = max(0, int(widthFactor * xRightTop))
        yRightTop   = max(0, int(heightFactor * yRightTop))
        if class_id in classNames:
            output.append([(xLeftBottom, yLeftBottom),
            (xRightTop, yRightTop)])
    return output
```

6.6: The main function of the *CharacterRecognizer*.

```
import tensorflow as tf
charset = {'0': 0, '1': 1, '2': 2, '3': 3, '4': 4, '5': 5, '6': 6,
           '7': 7, '8': 8, '9': 9, 'A': 10, 'B': 11,
           'C': 12, 'D': 13, 'E': 14, 'F': 15, 'G': 16, 'H': 17, 'I': 18,
           'J': 19, 'K': 20, 'L': 21, 'M': 22,
           'N': 23, 'O': 24, 'P': 25, 'R': 26, 'S': 27, 'T': 28, 'U': 29,
           'V': 30, 'X': 31, 'Z': 32, '<': 33,
           '>': 34, '(': 35, ')': 36, '$': 37, '#': 38, '^': 39, 's': 40,
           '-': 41, '*': 42, '%': 43, '?': 44, '!': 45, '+': 46}

# load frozen graph
graph_path = '~/models/ocr_graph.pb'
recognition_graph = tf.Graph()
with recognition_graph.as_default():
    od_graph_def = tf.GraphDef()
    with tf.gfile.GFile(graph_path, 'rb') as fid:
        serialized_graph = fid.read()
        od_graph_def.ParseFromString(serialized_graph)
        tf.import_graph_def(od_graph_def, name='')
session = tf.Session(graph=recognition_graph)
# prepare input and output request
output = []
```

```
gra_input = recognition_graph.get_tensor_by_name('ocr_input:0')
output.append(recognition_graph.get_tensor_by_name('predicted_chars:0'))
output.append(recognition_graph.get_tensor_by_name('predicted_scores:0'))
idx_lbl = {}
for key in charset.keys():
    idx_lbl[charset[key]] = key

def predict(image_np):
    assert image_np.shape == (180, 180, 3)
    img_in = numpy.expand_dims(image_np, axis=0)
    codes, scores = session.run(output, feed_dict={gra_input:
    img_in})
    codes, scores = [numpy.squeeze(x) for x in [codes, scores]]
    score_ave = 0
    text = ''
    for char, score in zip(codes, scores):
        if not idx_lbl[char] == '+':
            score_ave += score
            text += idx_lbl[char]
    score_ave /= len(text)

    return text, score_ave
```

Robot System Integration

7

Nguyen Van Toan

In previous chapters, a robot system is proposed to work in modern multi-floor buildings as customers' requests. Besides, a framework for the proposed mobile manipulator to change its work-floors is also presented, which is used as a particular example to demonstrate the final goal of the book without loss of generality. Then, single robotic modules are conducted to complete individual missions related to the proposed framework, including mobile robot navigation, manipulator manipulation, and robot perception works. In this chapter, a robotic decision-making system is presented to integrate single robotic modules, whose purpose is to help the proposed robot system switch its missions automatically, without human interventions.

ROBOTIC DECISION-MAKING SYSTEM

To be compatible with presented robotic modules, the robot system integration is conducted on the robot operating system (ROS), which is a middleware framework, including a set of open-source libraries and tools to build robot applications. In ROS, finite state machines (FSMs) and Behavior Trees (BTs) are used to model complex robot behaviors and multi-step tasks. Here, FSM is a computational model based on a finite number of states of a system, which has been used for many years as a task-switching architecture [1–9]. The concept of state is familiar and is famously illustrated by a simple mechanism of a drop-arm turnstile, as presented in Example 11.27 (page 762, Chapter 11) in [3]. This drop-arm turnstile is a form of gate with some rotating arms at waist height, which allows one person to pass at a time by inserting a coin (or other methods of payment). At first, its arms

DOI: 10.1201/9781003352426-7

are locked to prevent persons from passing through. To pass through the turnstile, the commuter must deposit a coin in a slot on the turnstile to unlock and allow its arms to rotate. Two possible states of the turnstile are *"locked"* and *"unlocked."* Two possible inputs that affect its states are *"deposit a coin"* and *"push the arms."* The input *"deposit a coin"* shifts the state of the turnstile from *"locked"* to *"unlocked."* In the *"unlocked"* state, it is possible to push the arms, and depositing additional coins does not change the state of the turnstile. However, the input *"push the arm"* will shift the state of the turnstile from *"unlocked"* to *"locked."* Once it is in the *"locked"* state, pushing the arms has no effect on the state of the turnstile. Through the lens of this example, an FSMs can be considered as an abstract machine that is defined by a list of expected *states* of the system, a set of potential inputs, and a set of probable outputs corresponding to potential inputs. At any given time, the FSM can be in exactly one of a finite number of *states*. The change from one state to another is called a *transition*, which is triggered by the input associated with it. Basically, FSMs are classified into two types: deterministic FSM and non-deterministic FSM; therein, a deterministic FSM can be constructed equivalent to any non-deterministic FSM. Over the years, event-driven approaches have been popularly developed for modeling the actions of an FSM in which the transitions are labeled by events [2]. These frameworks have some drawbacks as analyzed in [1] and [9]. Firstly, they are implemented as a set of concurrent threads which communicate with each other. Therefore, an additional process is required to manage the synchronization of the threads and ensure there is no deadlock or thread starvation. As a consequence, it increases the load on the system. Secondly, the idealized model of physical chance events, arranged in an infinitely timeline, can only be approximated in software. It causes nondeterministic memory usage and timing. So, it is not suitable for reactive real-time systems. Finally, the verification process can lead to a combinatorial explosion since all combinations of FSM states need to be correct. This significantly affects the formal verification of the models both in the time domain and in the value domain. This observation raises the need to propose alternative approaches which support single-threaded execution of multiple FSMs [5]. By way of explanation, FSMs are known as one-way control transfers, in which a trade-off between reactivity and modularity is occurred.

BTs are known as graphical mathematical models for reactive and fault-tolerant task executions, which were first introduced in the computer game industry [10]. Later, the BTs are popularly used in textbooks [11], learning [12], task planning [13], robotics [14–16], human-robot interaction [17], multi-robot systems [18], and system analysis [19]. Being motivated by the high modularity, flexibility, and reusability, the BT is interestingly emphasized on the Navigation Stack of ROS2 [20]. In BTs, internal nodes represent behavior compositions while leaf nodes represent actuations or sensing operations. Popularly, they are divided into root, parent, and child nodes. Here, a BT is executed from the root node. During the execution of a BT, a node in the tree is only executed if it received activation signals. The control transfer in the tree is described as follows: after the parent node received activation signals, it routes these signals to its child nodes with a given frequency. The child node receives activation signals and is executed. Then, the child node returns a status (Success, Running, or Failure, according to the logic of the node) to the parent. If activation signals are no longer received, the execution of the child node is stopped. As presented in [21, 22], the classical representation of BTs consists of four composition nodes (Sequence, Fallback, Parallel, and Decorator) and two execution nodes (Action and Condition). In Sequence, Fallback, and Parallel nodes, activation signals are routed to their child nodes if these signals are received. The status of Sequence node is Failure or Running if a child (C_i) returned Failure or Running (in this case, the Sequence node keeps sending activation signals to all the children up to the child C_i and stops sending activation signals to the child C_{i+1}) and is Success if all the children return Success. By contrast, the status of Fallback node is Success or Running if a child (C_i) returns Success or Running (in this case, the Fallback node keeps sending activation signals to all the children up to the child C_i and stops sending activation signals to the child C_{i+1}) and is Failure if all the children return Failure. In the different logic, the Parallel node returns Success if M child nodes return Success, returns Failure if more than N-M child nodes return Failure, and returns Running otherwise. Singularly, the Decorator node is used to introduce additional semantics or to change the return status of a node according to the custom-made policy, which represents a particular control with only one child. Action nodes perform operations corresponding to activation signals. An action node returns Success if the operations are completed, Failure if the operations cannot be completed, or Running otherwise. Similarly to other nodes, the execution of an Action node is aborted if it no longer receives activation signals. Finally, the Condition node is used to check whether a proposition is satisfied (return Success) or not (return Failure) after it received activation signals. By way of explanation, the state transition logic in BTs is organized in a hierarchical tree structure. As a result, the modularity of the system is extended. Unlike FSM, BTs are two-way control transfers that are controlled by the internal nodes of the tree. Therefore, it can manage states more efficiently [20]. The advantages of BTs are useful for the modern robot system integration. In [22], some improvements for the BT implementation are presented. Besides, BTs fit inside the robotic software architecture are also suggested, in which the robotic software is categorized in Mission Layer (defined higher level goals for the robot to achieve), Task Layer (defines how the robot accomplishes a goal, disregarding the implementation details), Skill Layer (defines the basic capabilities of a robot, describing the implementation of leaf nodes of a BT), and Service Layer (contains entities that serve as the access point for the skills to command the robot).

For a system with a small number of states, the difference in using an FSM or a BT for its execution seems to be negligible. However, the structure of the FSM will become much more complex than the BT if the number of transactions of the system is big. To understand more deeply in FSM, BTs, and their applications, it is encouraged to obtain more details in mentioned references.

AN EXECUTION OF ROBOT BEHAVIORS

In this section, an integration of single robotic modules is conducted to help the proposed robot system switch its missions automatically, without human interventions. Here, Tasks, Missions, and Actions of the proposed mobile manipulator are defined through presented mobile robot navigation, manipulator manipulation, and perception works. Firstly, the task list of the robot system should be formulated. To do this, an *xml* tree is used, as following the below structure:

- The XML tree is organized as: Task–Mission–Action.
- One Task includes various Missions (as sub-elements of Task), and one Mission has various Actions (as sub-elements of Mission).
- The space (" ") is used to separate input values of actions.
- The rule for Action parameters:
  - If there are two or more values, the comma (",") is used to separate them.
  - The structure "name:input" is used to input the name and the value of parameters.
- Mission names cannot be duplicated, while action names can be duplicated.

More details of *xml element tree* can be found in [23]. Based on the proposed working-floors-changing framework, an *xml* file is created and named as "*example_task.xml*"; its contents are as follow:

```
example_task.xml:
<task name = "Floor_Change_Task" loop="1">
```

```
<mission name="EV_MISISON1" loop = "1">
    <action name="waypoint" value="10.2 5.5 -1.2"
    param="max_trans_vel:0.5, max_rot_vel :0.5" />
    <action name="elevator" value="1" param=" " />
    <action name="elevator" value="2"
    param="current_floor:'2', target_floor: '4'" />
    <action name="elevator" value="3" param=" " />
    <action name="basic_move" value="0.0"
    param="distance:1.5" />
    <action name="basic_move" value="1.0"
    param="distance:1.57" />
    <action name="standby" value="3" />
    <action name="elevator" value="2"
    param="current_floor:'2', target_floor: '4'" />
    <action name="elevator" value="4"
    param="current_floor:'2', target_floor: '4'" />
    <action name="elevator" value="3" param=" " />
    <action name="basic_move" value="1.0"
    param="distance:1.57" />
    <action name="basic_move" value="0.0"
    param="distance:1.5" />
    <action name="standby" value="3" />
</mission>
</task>
```

Above, the parameter "*loop*" is used to define how many times the task or the mission is executed. Moreover, actions (*waypoint*, *elevator*, *basic_move*, *standby*) and their parameters are defined as below:

- *waypoint*: This action is to set target poses for the mobile robot. Therefore, its values and parameters are similar to values and parameters of a waypoint in the navigation mode. Some necessary values and parameters can be illustrated as follows:
  - values:

NAME	TYPE	DESCRIPTION
x	float	the x-coordinate of the targeted point
y	float	the y-coordinate of the targeted point
theta	float	the theta at the targeted point

  - parameters:

NAME	TYPE	DESCRIPTION
max_trans_vel	float	maximum translation speed

max_rot_vel	float	maximum rotation speed
xy_tolerance	float	target coordinates (x, y) tolerance
yaw_tolerance	float	target angle (theta) tolerance
max_trans_acc	float	maximum linear acceleration
max_rot_acc	float	maximum angular acceleration

- *elevator*: This action is to implement works related to the elevator operation. As presented, the elevator operation includes the elevator button detection, the button light-up checking, the door status checking, the human detection, and the LCD number recognition. Here, the elevator operation is separated into four modes. Since the elevator button detection and the elevator button light-up checking are used together for the elevator button manipulation, they are then defined in the same mode.
  - values:

NAME	TYPE	DESCRIPTION
mode	int	1: human recognition 2: elevator button recognition and manipulation 3: door status checking 4: floor number checking

  - parameters:

NAME	TYPE	DESCRIPTION
current_floor	string	The current floor number. ex) '2'
target_floor	string	The target floor number. ex) '4'

- *basic_move*: This action is to make the mobile robot translating a distance or rotating an angle, with respect to its current position. The *basic_move* action is used in some areas where the localization of the mobile robot is inaccurate or where the mobile robot cannot localize. For example, the *basic_move* action is used inside the elevator or before the mobile robot get out of the elevator to enter the new floor and switch its working map.
  - values:

NAME	TYPE	DESCRIPTION
mode	int	0: linear movements 1: angular movements

- parameters:

NAME	TYPE	DESCRIPTION
distance	float	Linear distance (m), under mode 0 Angular distance (rad), under mode 1

- *stand_by*: This action is to make the robot system wait for a while after it finished an action or before it starts a new action. This action can be considered as the system hibernation, and it does not have any parameters.

- value

NAME	TYPE	DESCRIPTION
time	float	waiting time (s)

So far, the task list of the mobile manipulator is defined by using the *xml* tree, in which some basic action values and parameters are as mentioned above. It is reasonable to add more values and parameters for an action if they are required. To execute defined tasks, the system must read the task list from the *xml* file. This work can be implemented by using a function *getTask*, whose content is briefly presented in Appendix 7.1. Now, to get missions, actions (values and parameters), one more step should be conducted, as shown in Appendix 7.2. Here, four action modes are defined as: *WAYPOINT = 0x00, STANDBY = 0x02, ELEVATOR_MISSION = 0x03*, and *BASICMOVE = 0x04*.

Until now, the definition of the task list (including missions and actions) and the *xml* tree reading method are presented. Next, the robot actions should be mentioned, as follows:

```
import act_elevator
import act_basicmove
import act_waypoint
import act_standby
```

Here, *act_elevator, act_basicmove, act_waypoint*, and *act_standby* include classes and functions to implement elevator perception works, basic move actions, waypoint actions under the navigation mode, and standby actions, respectively. Main executions of *act_elevator, act_basicmove, act_waypoint*, and *act_standby* are named as *elevator_mission, basicmove_mission, waypoint_mission*, and *standby_misison*, respectively. Then, the action table can be defined as:

```
actions = {}
actions[WAYPOINT] = act_waypoint.waypoint_mission
```

```
actions[STANDBY] = act_standby.standby_mission
actions[ELEVATOR_MISSION] = act_elevator.elevator_mission
actions[BASICMOVE] = act_basicmove.basicmove_mission
```

There are four modes inside the act_elevator, defined as: *HUMAN_DETECT = 0x01, EV_BUTTON_CLICK = 0x02, DOOR_STATUS_CHECK = 0x03,* and *EV_FLOOR_CHECK = 0x04.* For example, *actions [ELEVATOR_MISSION](HUMAN_DETECT)* can be used to call the human detection function. Besides, the action wrapper for importing the action list is conducted:

```
was_actions = {}
for k in actions.keys():
    was_actions[k] = action_wrapper(actions[k])

def action_wrapper(cls, func):
    def wrapper(*args, **kwargs):
        return func(cls, *args, **kwargs)
    return wrapper
```

In all actions, seven work states are defined to provide the feedbacks when an action is executing. They are: *BEFORE = 0x05, EXECUTING = 0x06, PAUSE = 0x07, RESUME = 0x08, SUCCEEDED = 0x09, CANCEL = 0x10,* and *FAILED = 0x11.* Now, the final work is to use an action server to execute the defined actions. More details of this action server can be found in [24]. The main routine of the execution function is briefly presented in Appendix 7.3.

In this section, a robot system integration is presented to integrate navigation, perception, and manipulator-manipulation modules to help the mobile manipulator change its work-floors without human interventions. For more details, an observation on mentioned references is necessary.

ABBREVIATIONS

ROS: Robot Operating System
FSM: Finite State Machine
BT: Behavior Tree
XML: An Inherently Hierarchical Data Format
LCD: Liquid-Crystal-Display

REFERENCES

[1] Hermann Kopetz. *Should responsive systems be event-triggered or time-triggered?* IEICE Transactions on Information and Systems, vol. E76-D, no. 11, pp. 1325–1332, 1993.

[2] David Harel, Amnon Naamad. *The statemate semantics of statecharts.* ACM Transactions on Software Engineering and Methodology, vol. 5, no. 4, pp. 293–333, 1996.

[3] Thomas Koshy. Discrete mathematics with applications. Academic Press, Cambridge, 2004. ISBN 978-0-12-421180-3.

[4] John E. Hopcroft, Rajeev Motwani, Jeffrey D. Ullman. Introduction to automata theory, languages, and computation, 3rd edn., Pearson, Boston, 2006.

[5] Vladimir Estivill-Castro, David A. Rosenblueth. *Model checking of transition-labeled finite-state machines.* In Software Engineering, Business Continuity, and Education. ASEA 2011. Communications in Computer and Information Science, vol. 257, pp. 61–73, Springer, Berlin-Heidelberg, 2011. https://doi.org/10.1007/978-3-642-27207-3_8

[6] Maksym Figat, Cezary Zielinski, Rene Hexel. *FSM based specification of robot control system activities.* In Proceedings of the 11th International Workshop on Robot Motion and Control, Wasowo Palace, Poland, 3–5 July 2017, pp. 193–198.

[7] Richard Balogh, David Obdrzalek. *Using finite state machines in introductory robotics.* In Lepuschitz, W., Merdan, M., Koppensteiner, G., Balogh, R., Obdrzalek, D. (Eds.) Robotics in Education. RiE 2018. Advances in Intelligent Systems and Computing, vol. 829, pp. 85–91. Springer, Cham, 2018. https://doi.org/10.1007/978-3-319-97085-1_9

[8] Mordechai Ben-Ari, Franceso Mondada. *Finite state machines.* In Elements of Robotics. Springer, Cham, pp. 55–61, 2018. https://doi.org/10.1007/978-3-319-62533-1_4

[9] Razeen Hussain, Teresa Zielinska, Rene Hexel. *Finite state automaton based control system for walking machines.* International Journal of Advanced Robotic Systems, 2019, doi:10.1177/1729881419853182.

[10] Daminan Isla. *Handling complexity in the Halo 2 AI.* In Proceedings of Game Developers Conference, 2005.

[11] Michele Colledanchise, Petter Ogren. *Behavior trees in robotics and AI: an introduction.* In Chapman & Hall/CRC Artificial Intelligence and Robotics Series, 1st edn., CRC Press, Boca Raton, FL, 2018.

[12] Bikramjit Banerjee. *Autonomous acquisition of behavior trees for robot control.* In 2018 IEEE/RSJ International Conference on Intelligent Robots and Systems (IROS), Madrid, Spain, 01–05 October 2018, pp. 3460–3467.

[13] Xenija Neufeld, Sanaz Mostaghim, Sandy Brand. *A hybrid approach to planning and execution in dynamic environments through hierarchical task networks and behavior trees.* In Proceedings of 14th Artificial Intelligence and Interactive Digital Entertainment Conference, vol. 14, no. 1, pp. 201–207, 2018.

[14] Francesco Rovida, Bjarne Grossmann, Volker Fruger. *Extended behavior trees for quick definition of flexible robotic tasks.* In Proceedings of IEEE/RSJ International Conference on Intelligent Robots and Systems (IROS), Vancouver, Canada, 24–28 September 2017, pp. 6793–6800.

[15] Dianmu Zhang, Blake Hanaford. *IKBT: solving symbolic inverse kinematics with behavior tree.* Journal of Artificial Intelligence Research, vol. 65, no. 1, pp. 457–486, 2019.

[16] Chris Paxton, Andrew Hundt, Felix Jonathan, Kelleher Guerin, Gergory D. Hager. *CoSTAR: introducing collaborative robots with behavior trees and vision.* In 2017 IEEE International Conference on Robotics and Automation (ICRA), Singapore, 29 May–03 June 2017, pp. 564–571.

[17] Nils Axelsson, Gabriel Skantze. *Modelling adaptive presentations in human-robot interaction using behavior trees.* In Proceedings of the 20th Annual SIGdial Meeting on Discourse and Dialogue, Stockholm, Sweden, 11–13 September 2019, pp. 345–352.

[18] Oliver Biggar, Mohammad Zamani. *A framework for formal verification of behavior trees with linear temporal logic.* IEEE Robotics and Automation Letters, vol. 5, no. 2, pp. 2341–2348, 2020.

[19] Petter Ogren. *Convergence analysis of hybrid control systems in the form of backward chained behavior trees.* IEEE Robotics and Automation Letters, vol. 5, no. 4, pp. 6073–6080, 2020.

[20] Steve Macenski, Francisco Martín, Ruffin White, Jonatan Ginés Clavero. *The Marathon 2: a navigation system.* In 2020 IEEE/RSJ International Conference on Intelligent Robots and Systems (IROS), Las Vegas, NV, USA, 25–29 October 2020, pp. 2718–2725.

[21] Matteo Iovino, Edvards Scukins, Jonathan Styrud, Petter Ogren, Christian Smith. *A survey of behavior trees in robotics and AI.* Robotics and Autonomous Systems, vol. 154, p. 104096, 2022.

[22] Michele Colledanchise, Lorenzo Natale. *On the implementation of behavior tree in robotics.* IEEE Robotics and Automation Letters, vol. 6, no. 3, pp. 5929–5936, 2021.

[23] The Element Tree XML API. Available at: https://docs.python.org/3/library/xml.etree.elementtree.html

[24] Action Server and Client. Available at: http://docs.ros.org.ros.informatik. uni-freiburg.de/en/rolling/Tutorials/Actions/Writing-a-Py-Action-Server-Client.html

APPENDIX

7.1: The function "*getTask*" to load the defined task list in the *.xml* file.

```
import xml.etree.ElementTree as elemTree
task_path = 'example_task.xml'

def getTask(task_path):
```

```
try:
        task_list = []
        tree = elemTree.parse(task_path)
        tasks = tree.getroot()
        for task in tasks:
                msg = []
                for mission in task:
                        data = []
                        actionlist = mission.findall('action')
                        for action in actionlist:
                                data.append(action.attrib)
                        mission.attrib['action'] = data
                        msg.append(mission.attrib)
                task.attrib['mission'] = msg
                task_list.append(task.attrib)
        return task_list
except Exception as e:
        return None
```

7.2: The function to read the missions and actions in the task list.

```
def getActionIndex(name):
        try:
                if name == "waypoint" or name == "WAYPOINT": return WAYPOINT
                elif name == "basic_move" or name == "BASICMOVE":
                return BASICMOVE
                elif name == "standby" or name == "STANDBY": return STANDBY
                elif name == "elevator" or name == "ELEVATOR":
                return ELEVATOR_MISSION
                else : return -1
        except Exception as e:
                return None
def task_read(task_list, _index):
        try:
                tasks = task_list
                index = _index
                task = tasks[index]
                task_name = ''; task_loop = 1
                if 'name' in task: task_name = task['name']
                if 'loop' in task:
                        try:
                                task_loop = int(task['loop'])
                        except Exception as e:
                                task_loop = 1
                if 'mission' in task:
                        for mission in task['mission']:
                                mission_name = ''; mission_loop=1
                                action_list = []
                                if 'name' in mission: mission_name = mission['name']
                                if 'loop' in mission:
                                        try:
                                                mission_loop = int(mission['loop'])
                                        except Exception as e:
                                                mission_loop = 1
                                if 'action' in mission:
                                        for action in mission['action']:
                                                action_name = ''; action_args=[];
                                                action_param=[]
```

```
        if 'name' in action:
        action_name = getActionIndex(action['name'])
        if 'value' in action:
        args = action['value']

        # Check the space " " to split values of
        variables in the args
        ''' CODE TO GET ACTION VALUES HERE '''

        if 'param' in action:
            args = action['param']
            args = args.split(",")

        # Check the value "," to split values of
        # variables in the args
        # Then, check ":" to get values of params
            for arg in args:
                data = arg.split(":")
                ''' CODE TO GET PARAM VALUES
                HERE '''

    except Exception as e:
        return None
```

7.3: The main routine to implement the task list.

```
def execute_cb_def(goal):
        is_paused = False              # This signal is from external
                                       # rostopics, rosservices or action
                                       # state return
        state_return = BEFORE

        # Initialization for a new goal
        _action_index = goal.action_start_idx
        _loop_flag = goal.loop_flag
        _loop_index = 0

        if len(goal.work) < _action_index:
                node.get_logger.info("action index is out of the range…")
                return

        # Start the mission routines
        while rclpy.ok():
            while _action_index < len(goal.work):
                _action = goal.work[_action_index]
                # Got PAUSE signal
                while is_paused:
                    if cancel_signal():
                        is_paused = False
                        '''cancel the current action'''
                        return
                # Got CANCEL signal
                if cancel_signal():
                        is_paused = False
                        '''cancel the current action'''
                        return

            if _action.action_type == WAYPOINT:
```

```
            state_return = was_actions[WAYPOINT](_action)
    elif _action.action_type == STANDBY:
            state_return = was_actions[STANDBY](_action)
    elif _action.action_type == ELEVATOR_MISSION:
            state_return =
            was_actions[ELEVATOR_MISSION](_action)
    elif _action.action_type == BASICMOVE:
            state_return = was_actions[BASICMOVE](_action)
    else:
            node.get_logger.info("[%d]Got Wrong
            Action"%_action_index)

    if state_return == FAILED:
            if cancel_signal():
                    is_paused = False
                    '''cancel the current action'''
                    return
            is_paused = True

    # Action Index Increasing
    if not is_paused:
            _action_index = _action_index + 1

# Count the looping states
_loop_index = _loop_index + 1
_action_index = 0
if _loop_flag == 0: continue
elif _loop_flag == _loop_index: break
```

Index

Note: *Italicized* and **bold** pages refer to figures and tables, respectively.